Easy Ideas 1

Airplanes 2

Cars 3

Computers 4

Smartphones 5

Food 6

Nature 7

Space 8

Light 9

AI 10

STEM-Zen Program

Everyday Objects Bonus

Easy Ideas From Concepts to Critical Thinking 1	**Airplanes** From Four Forces to Flights 2	**Cars** From Actions to Autos 3
Computers From Digital to Data 4	**Smartphones** From Calls to Global Connects 5	**Food** From Eats to Energies 6
Nature From Atoms to All Life 7	**Space** From Elements to Us 8	**Light** From Suns to Sapiens 9
AI From Machine Muscles to Minds 10	**STEM-Zen Program** From Empty to Science EnLights	**Everyday Objects** From Ideas to Daily Items Bonus

Airplanes

Table of Contents

2) Airplanes
—Teacher Guide—

2) Airplanes
— Four Forces

2) Airplanes
— Past and Present

2) Airplanes
— Wedge Tools

Bonus: Wing Ways

2) Airplanes

—Teacher Guide—

Science

2) Airplanes

With airplanes, our feet are no longer gravity-glued onto land. We earthlings soar upwards to become sky-lings.

Science of AIRPLANES

Airplanes
Table of Contents

2) Lift

4) Drag

1) Thrust

3) Gravity

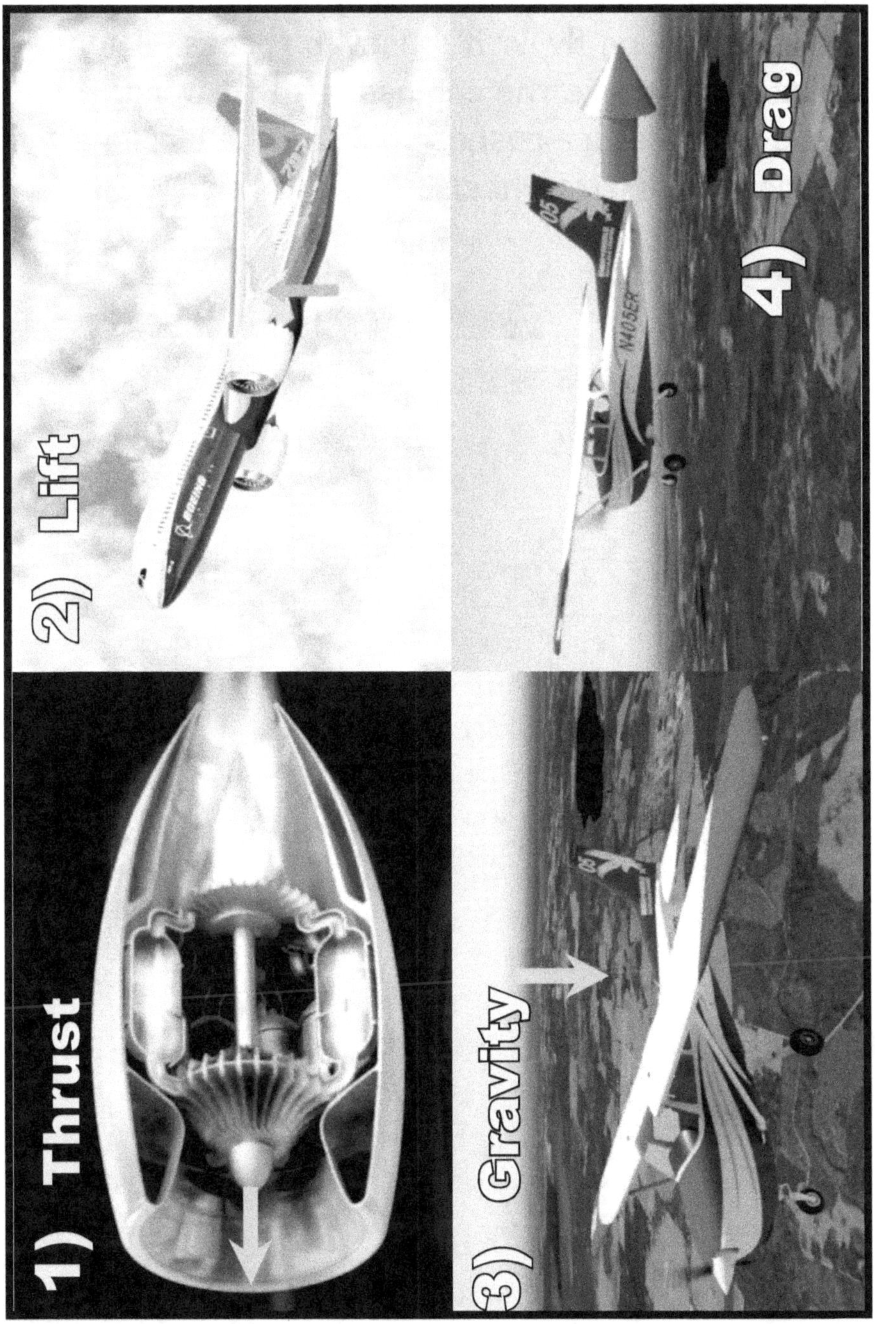

B) Roadmap

Purpose: Airplanes that weigh as much as 100 elephants can fly with 4 uneven forces. There is a point where the engines push forwards and air pressure pushes up when the airplane's weight is off its wheels. We call this "WOW."

Main Points
Airplanes with 4 Uneven Forces

1) Engines, <u>THRUST</u> forward.
2) Wings <u>LIFT</u> up.
3) <u>DRAG</u> pushes back.
4) <u>GRAVITY</u> pulls down.
5) Flight <u>controls</u> move the plane.
Rudder moves it left or right (yaw).
Elevators move it up or down (pitch).
Ailerons move it side to side (roll).
6) Pilots find their way or <u>navigate</u> airplanes. They plan and direct the plane's route using radar, GPS, sensors like accelerometers and gyros and other instruments like radio beacons on the ground.
7) Pilots use radios, which use radio waves, to <u>communicate</u> with people on the ground. Airport control towers are one example.

TEACHER PREP
C) <u>Refresh</u> KEY CONCEPTS
— Airplanes

AIRPLANES fly with the imbalance of four forces.

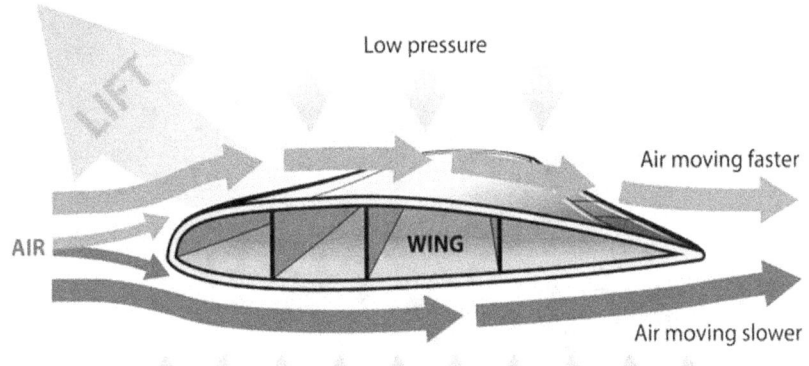

1) THRUST — is to push forward.
Jet fuel burns inside the engine. Exhaust pushes
backwards and thrusts the plane forwards.

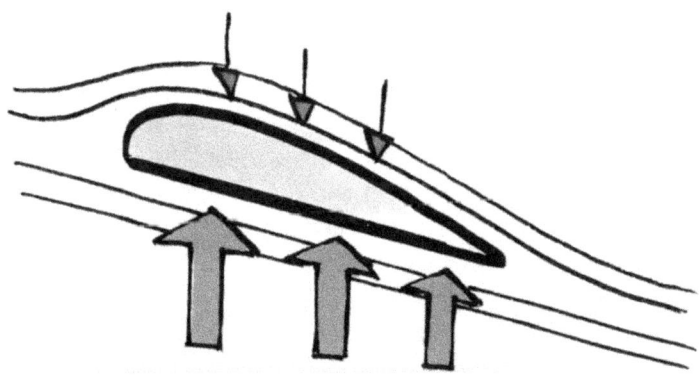

2) LIFT — is to push up.
Engines push air. Air flows over and under the wings
at different speeds. This causes greater pressure
under the wings that push the plane up.

3) GRAVITY — pulls the plane down.
Lift overcomes gravity to get the plane off of the ground. To land the plane, engines push less. Wings lift less and let gravity pull the plane back down.

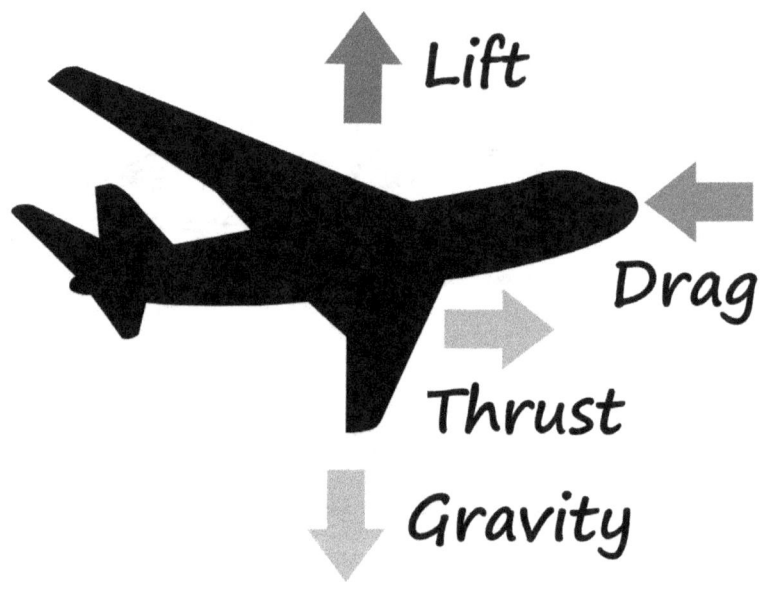

4) DRAG — pulls back and slows down the airplane.
It is the push back from the air in front. Streamline shapes help reduce drag. During landing, the shape of the wing changes to increase drag to slow the plane down.

Airplanes cruise when lift up equals gravity down and thrust forward is greater than drag back.

For a copy of this video contact:

When you watch this video, keep
in mind the four forces of flight.

1) Thrust Forward
2) Lift Up
3) Drag Back
4) Gravity Down

E) Ready / Present In-Person or On-Line Class.

Airplanes

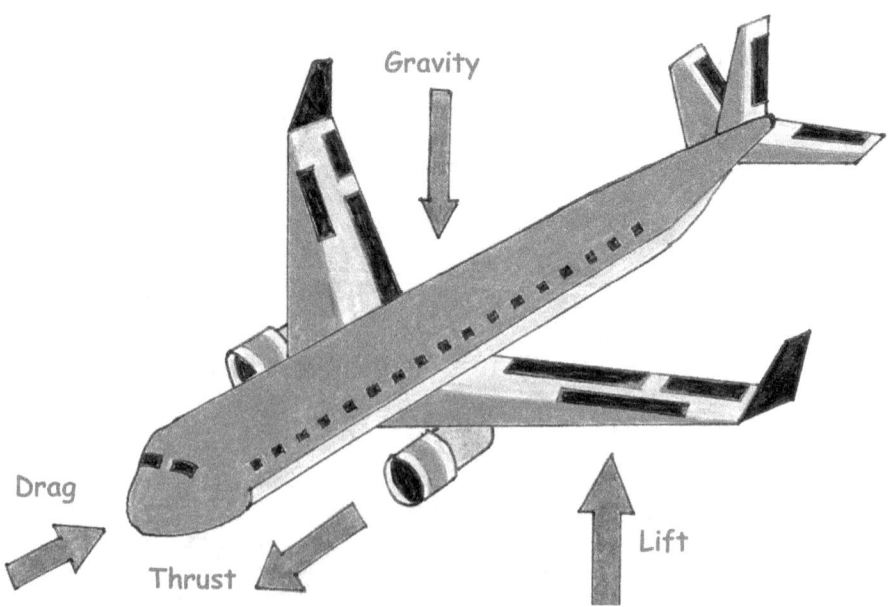

. Get the materials for
the "Do" Demonstrations.
— Paper Lift
— Aluminum Plane

. Printout "Try It" Worksheets.
— 1) Why Do Airplanes Fly?
— 2) What are Flight Controls?
— 3) What 3 Parts Control Plane?
— 4) Why Swept Wings?
— 5) Why Comets Crash?
— 6) How Much Weigh?

Gravity

Drag

Thrust

Lift

Planes fly with four uneven forces.

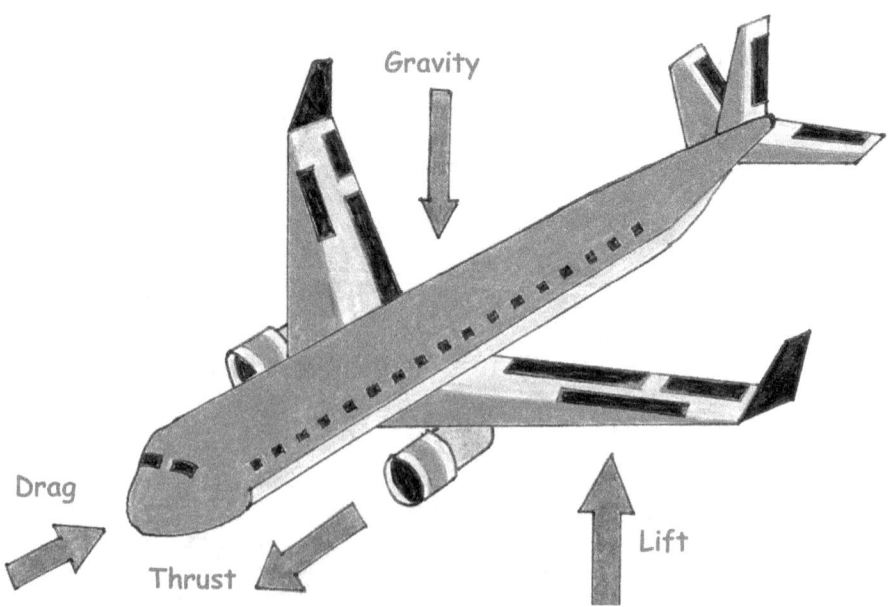

2) Airplanes
— Four Forces

Teachers use this eBook to teach students
about the four forces of flight:
lift, gravity, thrust and drag.

STEM-Zen Program

Four Forces

1) Thrust Forward

Engines push the plane forward.
There are two types of engines.

2) Lift Up

The shape of the wing forces air to flow
faster over the top. There are two ways to
think about how wings push or lift an airplane
up. The faster air from on top flows over
the wing and then down to push the plane up.

3) Drag Back

Air

Bike riders streamline their shape
so they go faster through the air.
Air in front of planes also pushes back.
This slows the plane and is called drag.
During flight, the engines have to
push harder to overcome drag.
Drag is useful to slow the plane
down when you want to land.

4) Gravity Down

A large object
pulls on smaller
objects. This
is called gravity.
Gravity pulls
apples down
from trees.
Gravity also
keeps us on
Earth.

When a plane is flying, slowing
the plane leads to less lift.
Gravity then pulls the plane down.

See the eBook for more about the 4 forces of flight.

Teacher - Airplanes

— Paper Lift

Purpose - Show how air lifts wings.

Materials. - A4/letter paper
- air

Steps: 1) Hold a piece of paper by the
long end with two hands as shown.
2) Blow over the top of the paper.
3) The air on top moves faster and
has less pressure then the air below.
4) Air below has higher pressure
and pushes or lifts the paper up.
5) Explain that the shape of the wing
is important to why planes fly.

DO!

— Aluminum Plane

Purpose - Show how aluminum planes fly.

Materials. - A4/letter paper
- aluminum foil
- scotch tape

Steps.
1) How do aluminum planes fly?
2) Wad up a sheet of Aluminum foil.
   Let it fall to the floor. Ask students
   to notice the shape of wadded sheet.
3) Tape sheet of foil onto the paper.
4) Fold into an airplane.
5) Fly
6) Explain that the shape of the wing
   is important to why planes fly.

Related Topics. Engine Thrust, Lift, Gravity

Question?

1) In four words, why do airplanes fly?

1) _____
   _____

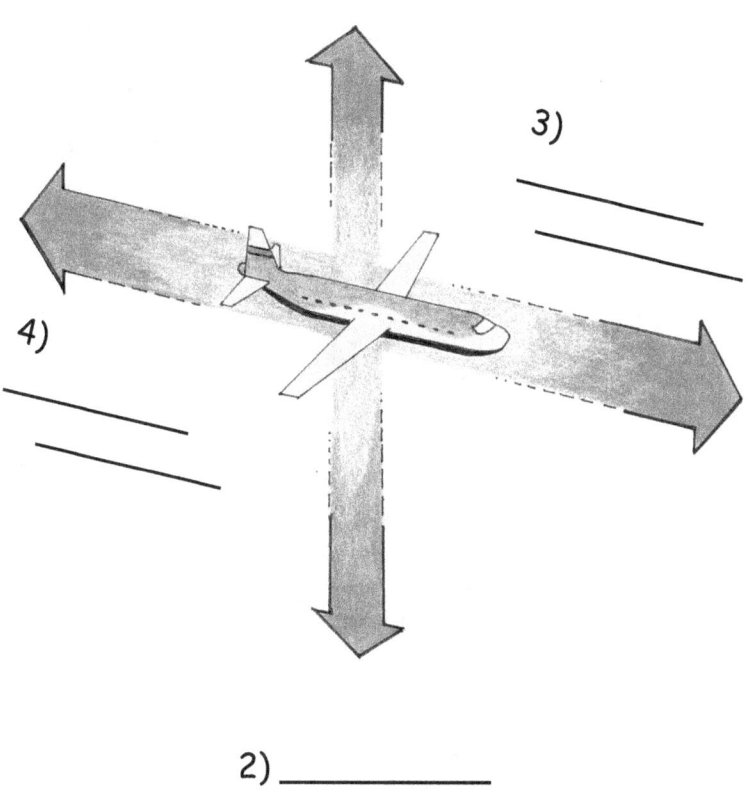

3)
   _____
   _____

4)
   _____
   _____

2) _____
   _____

TRY IT!

Question?

2) What are three Flight Control directions called?

3) _____

1) _____

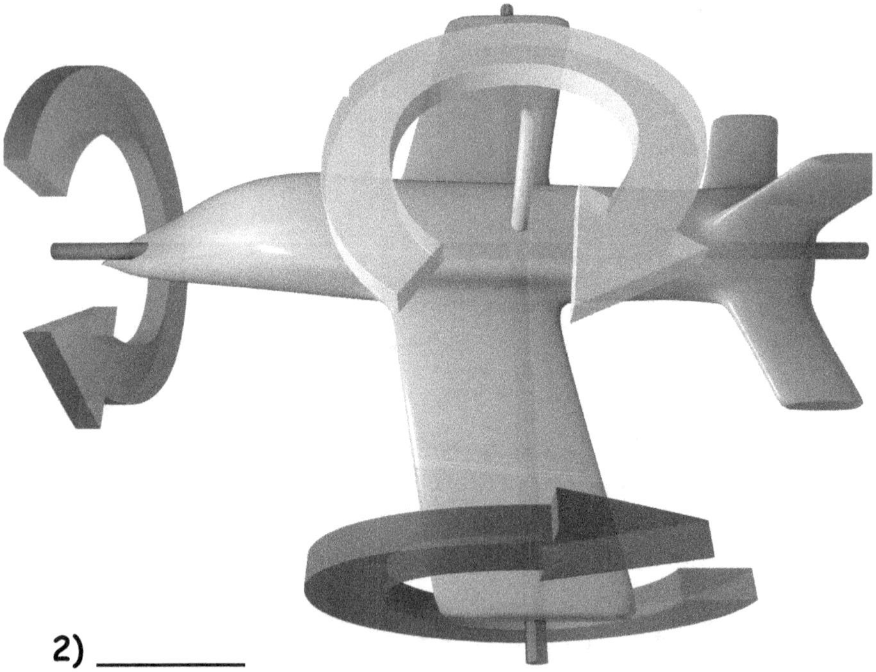

2) _____

Question?

3) What are the three parts that control the plane?

1) _____

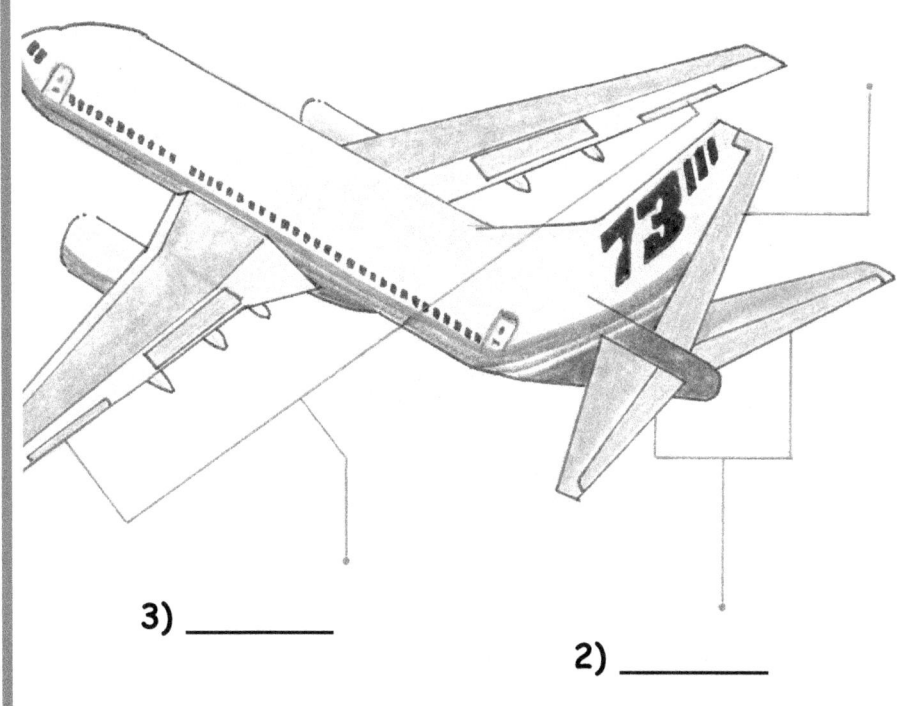

73"

3) _____

2) _____

Question?

4) Why do jet airplanes have swept wings and engines hanging below?

_____

_____

_____

_____

_____

_____

Question?

5) Why did the first commercial jet aircraft called "Comets" crash?

_____
_____
_____
_____
_____
_____

Question?

6) How much does a wide body commercial aircraft weigh?

TRY IT!

Answers

1) In four words, why do airplanes fly?

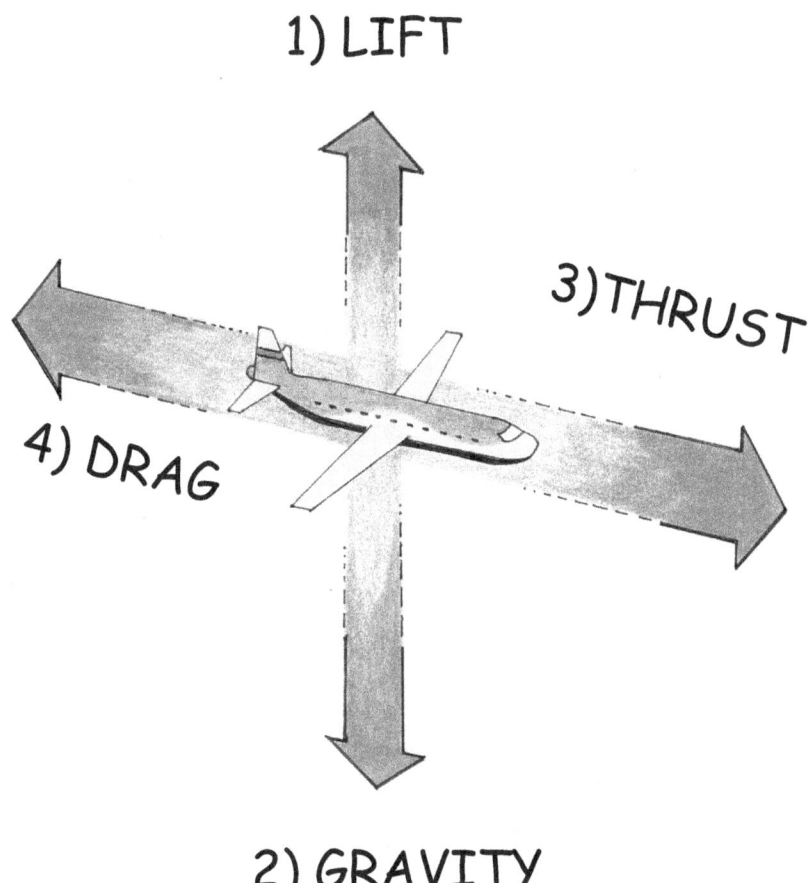

1) LIFT

3) THRUST

4) DRAG

2) GRAVITY

TRY IT!

Answers

2) What are three Flight Control directions called?

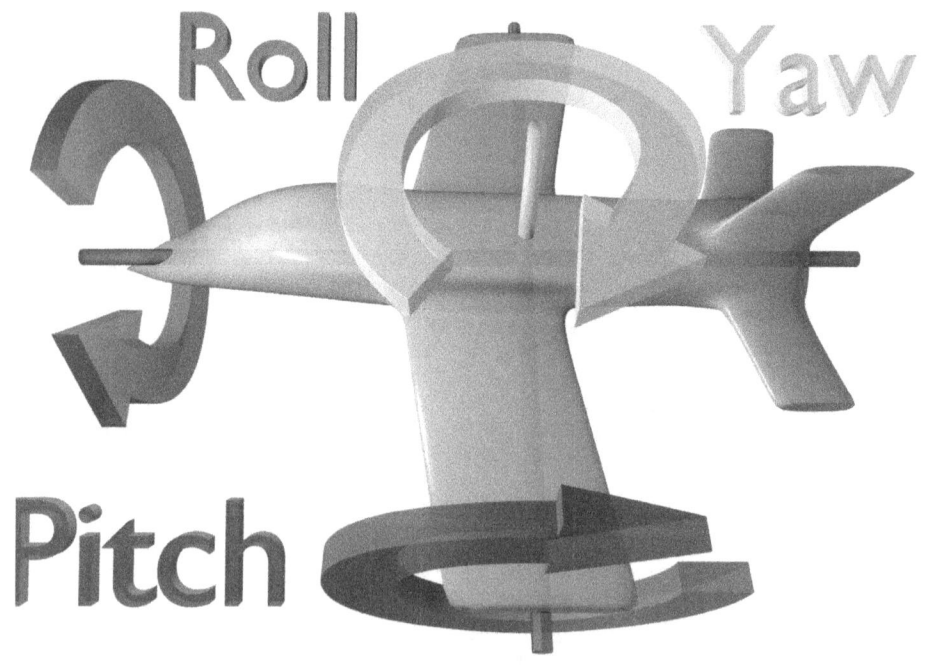

1) Yaw
2) Pitch
3) Roll

TRY IT!

Answers

3) What are the three parts that control the plane?

1) Rudder

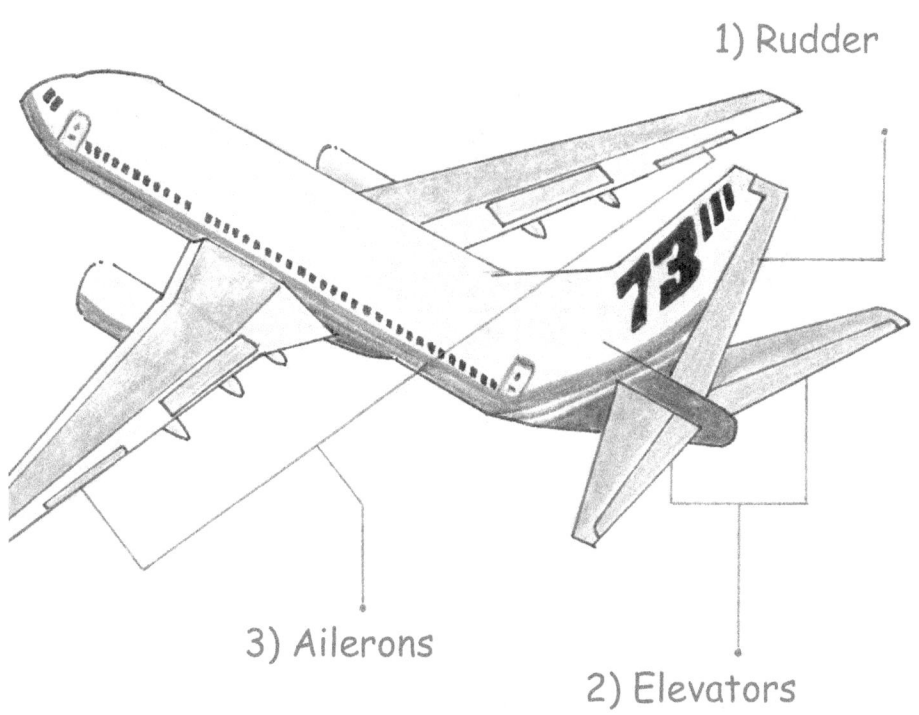

3) Ailerons

2) Elevators

TRY IT!

Answers

4) Why do jet airplanes have swept wings and engines hanging below?

Propeller planes have straight wings. Jet engines are so strong they can cause a problem called "flutter." Flutter is uncontrolled bending up and down of the wing. This can break the plane. Swept wings with engines below the wing, fix this problem.

TRY IT!

Answers

5) Why did the first commercial jet aircraft called "Comet" crash?

Metal fatigue caused cracks in the fuselage at the corners of the **square** windows. This caused the planes to crash. Today, planes have **round** windows.

TRY IT!

Answers

6) How much does a wide body commercial aircraft weigh?

 **x 100**

At take-off, a fully loaded wide-body airplane weigh as much as one hundred elephants.

F) RECAP
Airplanes with 4 Uneven Forces

1) Engines, THRUST Forward.

2) Wings LIFT Up.

3) DRAG pushes Back.

4) GRAVITY pulls Down.

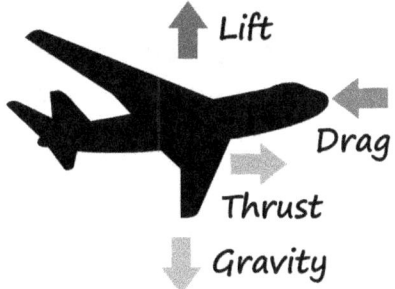

5) Flight controls move
the plane. Rudder moves it left or right (yaw).
Elevators move it up or down (pitch).
Ailerons move it side to side (roll).

6) Pilots find their way or navigate airplanes.
They plan and direct the plane's route using
radar, GPS, sensors like accelerometers and
gyros and other instruments like radio
beacons on the ground.

7) Pilots use radios, which use radio waves,
to communicate with people on the ground.
Airport control towers are one example.

G) <u>ROLL-UP</u> / Integrate Science
Airplanes, Four Forces

Air is important to why planes fly. The
air all around our planet is called an "atmosphere".
Animals and people breathe air. The sun shines
unevenly on earth. This makes air to have
different amounts of heat or temperature.
This causes our weather. We drive our
cars in all types of weather.

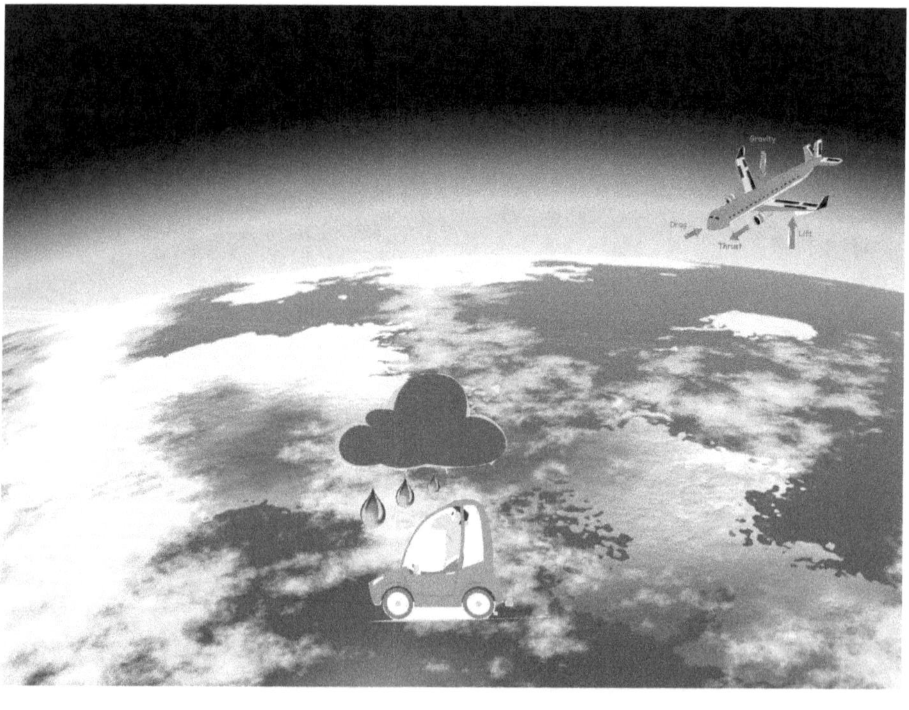

Advanced
— Jet Engines

The first airplanes were powered by gas engines with pistons that turned the propellers.
Next, jet engines work like this. Air enters the engine. The air is compressed, that is, squeezed together so more air fits into the engine. Air is mixed with fuel and burns or combusts. The hot exhaust pushes back and thrusts the plane forward.

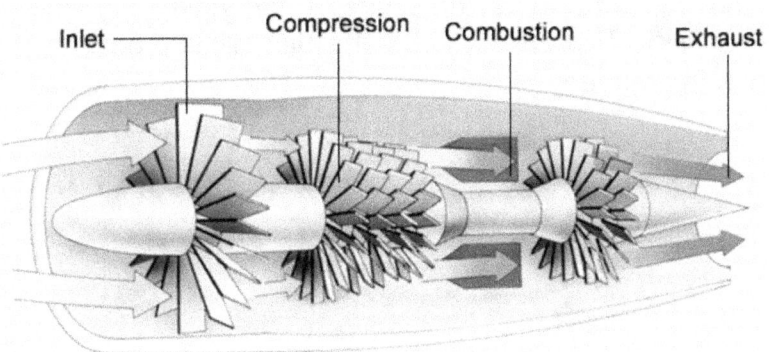

Inlet — Compression Combustion Exhaust

1) **Air Enters**
2) **Compresses**
3) **Combusts**
4) **Exhausts**

Science Story — People Glide

We start before airplanes are invented. In the 1800s,
Sir George Cayley notices birds gliding in the air at a windy
beach. He wants to make a glider for people. He makes
test equipment and does experiments to learn about wing
lift at different angles. Over decades he designs gliders
for people.

There are two versions of the next part of the story.
Sir George's horse-pulled carriage chauffeur climbs
in a glider at the top of a hill. The glider with the
chauffeur inside glides in the air.

At the bottom of the hill, the shaken but unhurt
chauffeur gets out of the glider. He marches right
over to his boss and quits his job. He said, "I am hired
to drive your carriage, sir but not fly in your inventions,
I must respectfully resign my position."

In another version of the story, it is a young boy who
glides first. Either way, it is the first time a person
glides in a glider.

Science Story — Toy Take Offs

In the 1880s a father gives his two young sons a rubber-band powered flying toy. They are fascinated with flight from this point on. As adults, the Wright Brothers start with the gliders that others design. Next, they build a wind tunnel to test out their own different wing designs.

The Wright Brothers have a gas-powered engine they make in their bicycle company machine shop. They put it all together and go to Kitty Hawk Beach. Wind at the beach gives extra lift. Also, to save weight, their creation does not have wheels so the sand will come in useful when they land.

In 1903, the Wright Brothers successfully fly the first self-propelled airplane, optimistically called "Flyer." As a side note, the Wright Brothers control X, Y and Z 3-D movements by moving their hands, arms and hips. The first pilot lies down to fly. Also, there are no interiors on the first planes. Everything is open to the elements (weather).

So, people notice birds glide at beaches. Then others design and make gliders. Later, after wind tunnel tests and a lot of trial-and-error experiments, the Wright Brothers show us that people can fly.

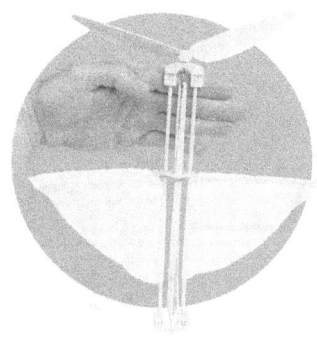 

Integrate

The first planes use piston-pushed propellers.
Decades later jet engines push airplanes forward.
To make a rocket we take the engine and turn it
90 degrees so it pushes up instead of over. We
have to add its own oxygen source to burn with
the fuel because there is no air in outer space.
Sixty-six years after the first airplane flies,
people walk on the Moon

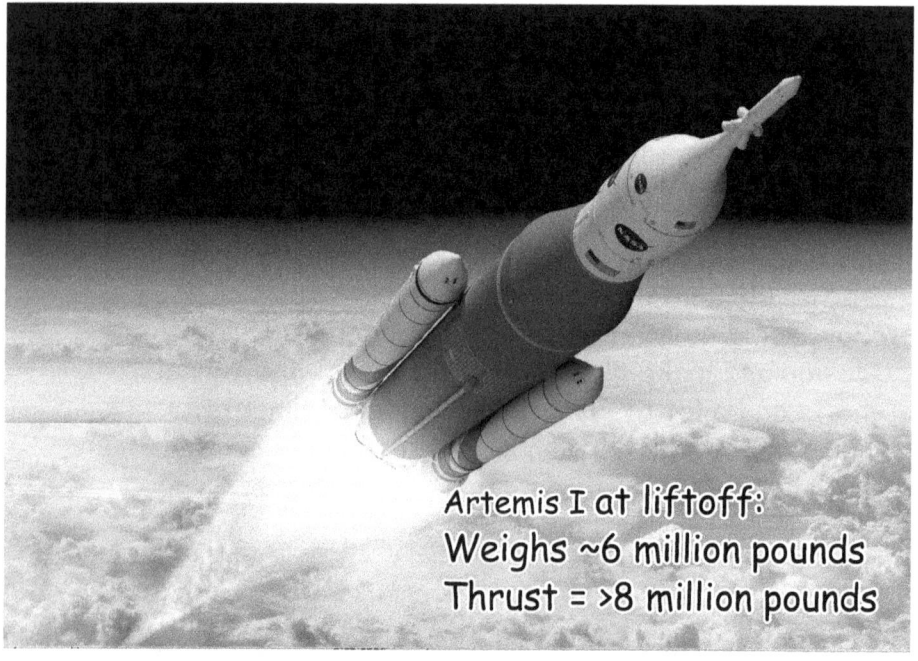

Artemis I at liftoff:
Weighs ~6 million pounds
Thrust = >8 million pounds

Science
of **Airplanes**
Four Forces

Planes
— Past &
Present

Wedge Tools
— Axes to
Airplanes

Read the ONE PAGER
Planes — Past and Present

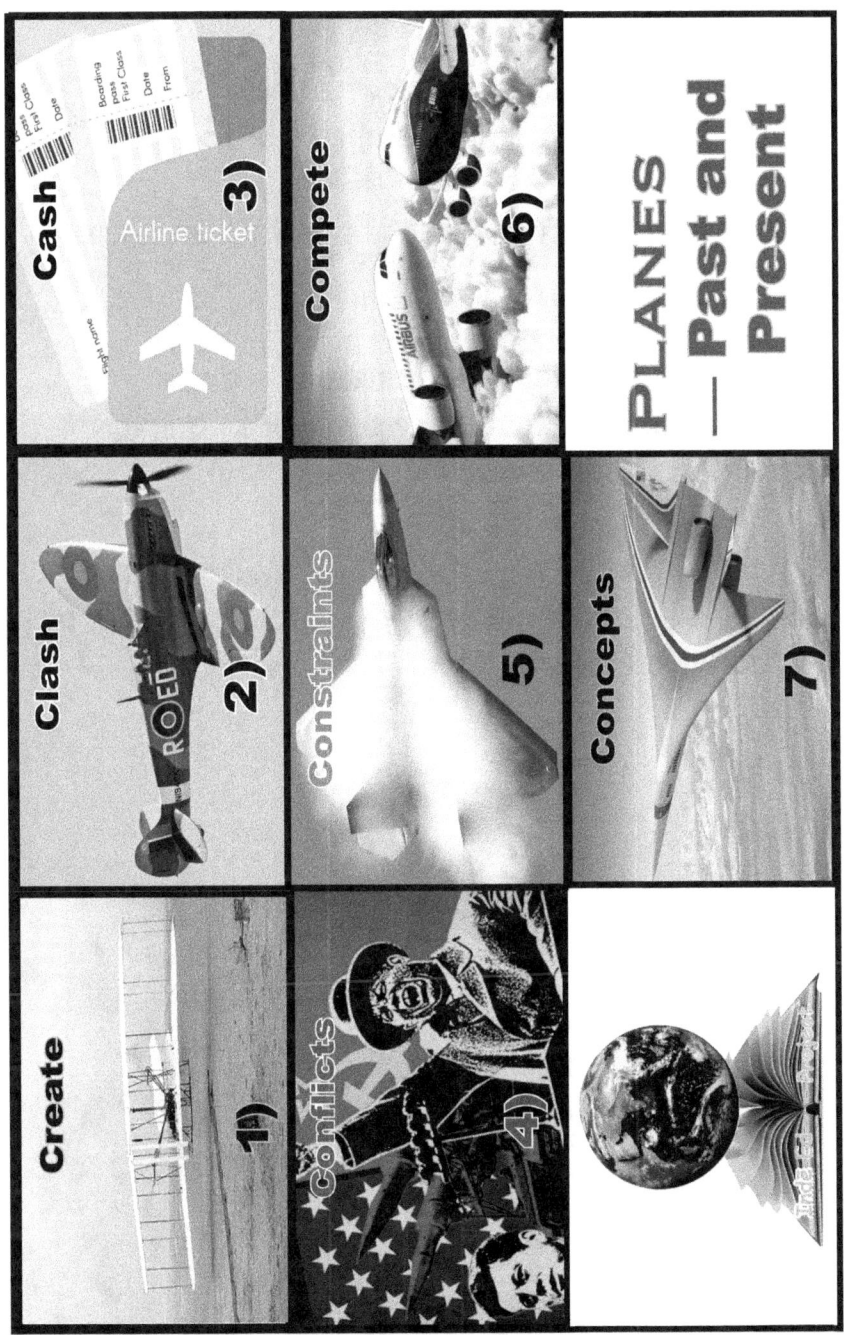

Planes — Past and Present

Purpose: Airplanes are awesome!
With science we humans are able to fly!

It takes 3 control surfaces and 4 forces to fly!

The first airplanes are controlled by:
. One hand moves the plane up or down
. Second hand moves the plane left or right
. The pilot is lying down and uses hips
   to tilt the plane side to side.
Human ingenuity continues to improve airplanes.
Sixty-six years after the airplane's first flight,
people walk on the Moon.

Main Points

1)   CREATE
People see birds and bugs fly.
People with science make the first airplanes.

2)   CLASH
People in different countries
compete to improve airplanes.

Main Points — continue
— Planes — Past and Present

3)　CASH
People are motivated by possible profits to make
airplanes that "quickly" fly paying passengers and
products around the world.

4)　CONFLICTS
Many aerospace advances are the result of airplanes
as weapons of wars.

5)　CONSTRAINTS
Planes with propellers can only go so fast due to
technical constraints. People invent jet engines to
fly faster and farther.

6)　COMPETE
Different companies & countries compete to make
airplanes more efficient with greater capabilities
like flying faster than the speed of sound.

7)　CONCEPTS
Today, people continue to think up concepts for
better airplanes. For example, electric engines,
the ability to fly to the edge of space and back
and imagining merging airplanes and flying cars.

Watch VIDEO

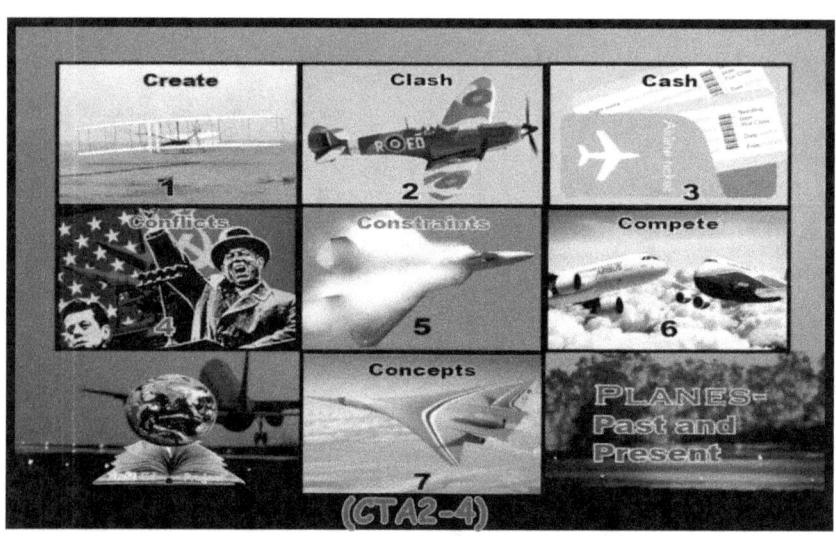

For a copy of this video contact:

<u>Main Points</u>
1) Create first flights
2) Flying machine clashes
3) Pay cash to fly
4) Cold War conflicts
5) Speed constraints
6) Countries &
companies compete
7) Future of flight concepts

Science Story
— Creative Flight Controls

How did the Wright Brothers control flight?
The first airplanes are controlled by:
. One hand moves the plane up or down (elevator).
. Second hand moves the plane left or right (rudder).
. The pilot is lying down and uses hips to roll the plane side
to side. Wires and pulleys actually twist one side of the
wing up and the other down depending on which direction
to roll the plane. This is called "wing warping."
Today, airplanes use ailerons to move side to side.

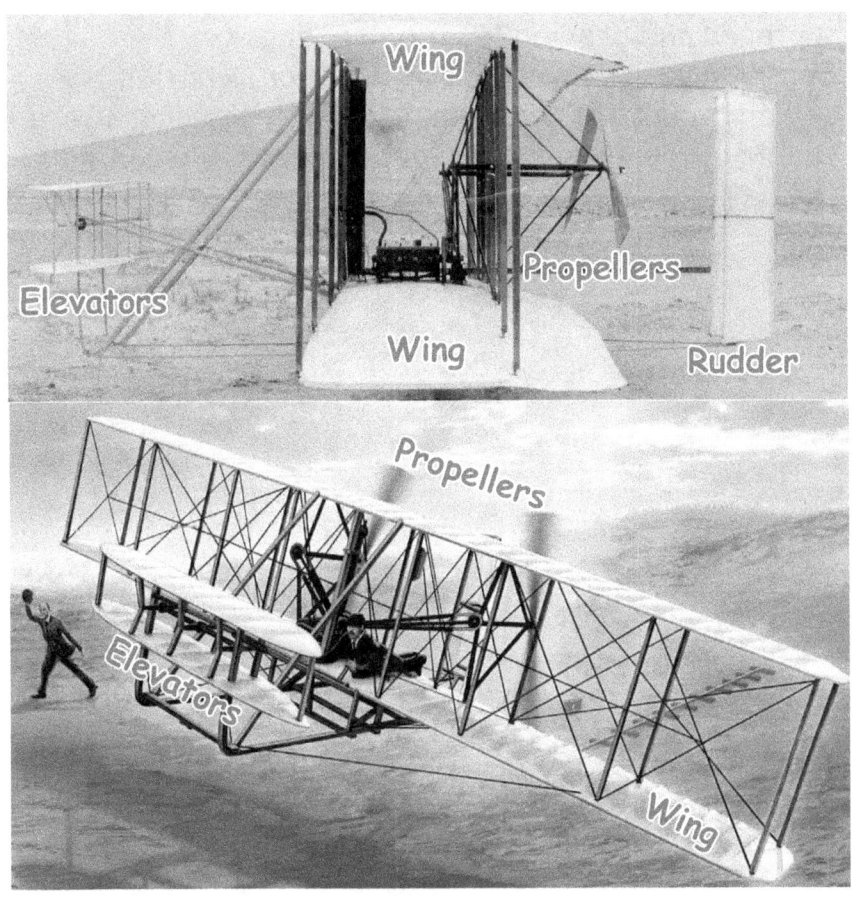

Advanced
<u>Planes — Past and Present</u>

How do modern planes do flight controls?
The rudder moves the plane left or right in the sky.
On the ground the steering wheel moves the plane
left or right. During flight, elevators in the tail move
the plane up or down. Ailerons roll the plane from
side to side. Pilots often combine the flight controls
to smoothly change the plane's direction.
Today, most planes are "fly-by-wire." The pilot
moves controls in the cockpit that are connected
by wires to flight management computers that are
linked to actuators that actually move the control
surfaces and provide location feedback.
Sensor inputs of aircraft orientation, location
and altitude that are connected to flight computers
enable autopilot software and hardware to control
the airplane.

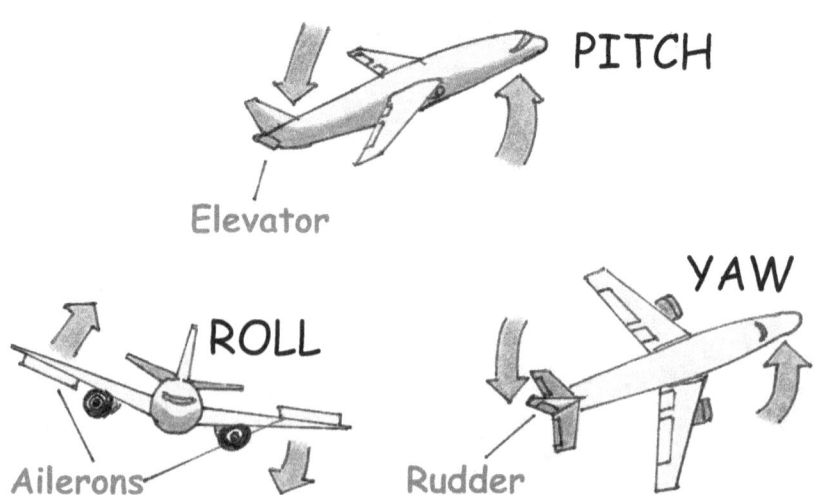

PITCH

Elevator

ROLL

Ailerons

YAW

Rudder

It is amazing that birds and bugs can fly. Notice how their wings
give thrust like an airplane engine and also lift. Think of the
powerful muscles that power animal flight. Equally awesome is how
their brains process input from senses and control their flight.

Advanced
Planes — Past & Present

Students compare and contrast the first propeller-powered Flyer and jet engine airplanes.

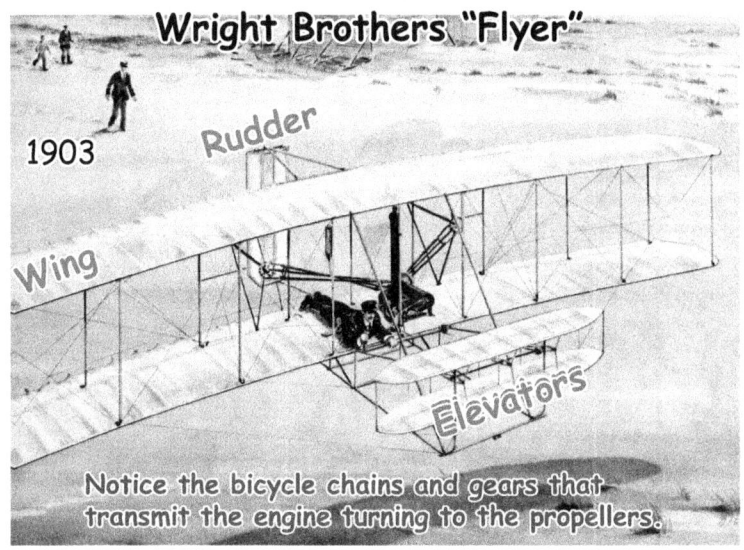

Wright Brothers "Flyer"

1903

Rudder

Wing

Elevators

Notice the bicycle chains and gears that transmit the engine turning to the propellers.

Modern Propeller Plane

Spoiler

Rudder

Elevators

Aileron

Spoilers flap up when the plane lands to increase drag and slow the plane down.

Integrate
Mars Shoot Goal

For perspective, the internal combustion gas engine is
invented about 30 years before the Wright Brothers
use it on their airplane. First flight is in 1903!
Sixty-six years later, in 1969, people walk on the Moon.
Can we, together, set the collective goal for humanity
to have settlements on the Moon and Mars 2035?
This is 66 years after our first moonwalk.

WEDGE TOOLS

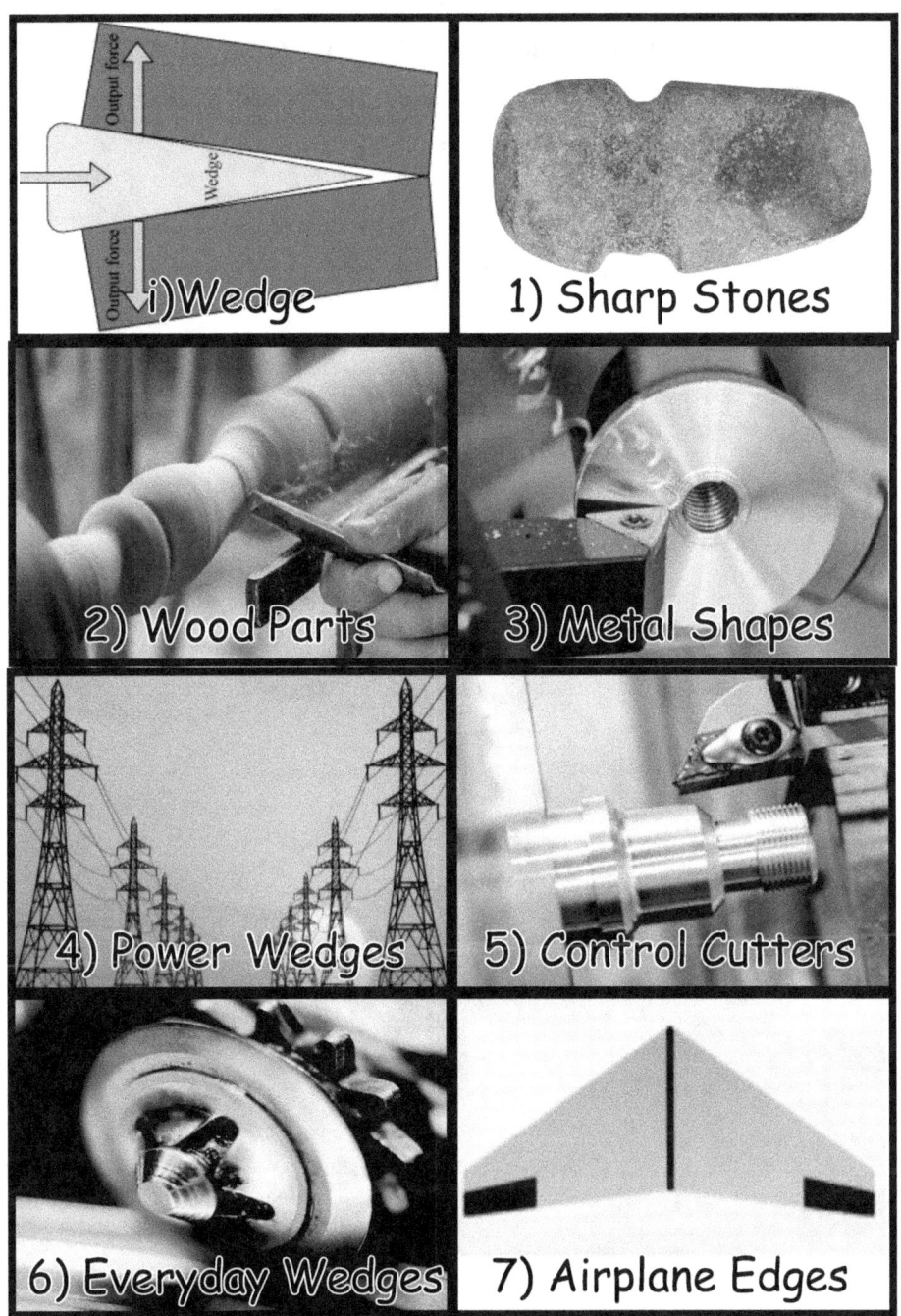

i)Wedge

1) Sharp Stones

2) Wood Parts

3) Metal Shapes

4) Power Wedges

5) Control Cutters

6) Everyday Wedges

7) Airplane Edges

Wedge Tools — Axes to Airplanes

Purpose: Science connects through time and technology to underpin our everyday objects. The link between wedge-shaped tools and swept-back jet wings is an example.

Main Points

i) WEDGES are V-shaped, triangle tipped tools.

1) SHARP STONES
Ancient people shape stones into V-shaped axes & arrowheads.

2) WOOD At first, wedge-tipped stone tools shape wood into useful things like houses, boats and fuel for fires.

3) METAL TOOLS Over time, people discover metals and turn them into tools like copper chisels to build pyramids and bronze weapons. Much later, iron and then steel world-changing wedge tools are made.

4) POWER People harness power to move triangle tipped tools. Over time, power sources are water, wind, steam and later electric and gas motors.

5) CONTROL Today, computers control machines that move cutting tools that make objects like cars and airplanes.

6) EVERYDAY Wedge-shaped objects are all around us like can-openers, scissors, razors, zippers, knives & tools.

7) WINGS Passenger airplane wings are wedge-shaped and swept-back. Leading edges are wedge-shaped too. So important are wedges that we call new high-tech "cutting edge."

Watch VIDEO
Cutting Wedges

For a copy of this video contact:

<u>Main Points</u>
i) Wedges
1) Sharp Stones
2) Wood Parts
3) Metal Shapes
4) Power Wedges
5) Control Cutters
6) Everyday Wedges
7) Airplane Edges

Advanced
Wedges — Axes to Airplanes

Science has synergy. That is, when parts are put together the result is greater than the sum of the pieces.

For example, a bag of parts doesn't do much until they are assembled to make a smartphone.

Ancient people shaped stone wedges into axes, spearheads and arrowheads. Next, people invented wedge-edged plows to prepare the land for planting crops. People use wedge-tipped sickles to harvest the grains. Over time people invent machines to move wedge-tipped cutting tools to produce parts for humans. Today, computers control the V-tipped tools to make modern objects like engines, shafts and gears for cars.

Science of Airplanes Four Forces — Planes — Past & Present — Wedge Tools — Axes to Airplanes

Science Story — Wedge Write

Cutting edges enable the world-wide Agricultural
Revolution. Our ancestors stop hunting and
gathering. They become farmers and city citizens.
At first, they use V-tipped planting sticks. Next,
they create wedge-edged hand tools like hoes.
They invent animal pulled wedge-edged plows
that prepare the soil for seed planting.
In the first cities in the Fertile Crescent the first
known writing is invented. It uses wedge-shaped
characters called "cuneiform."
We know about this because lots of sunbaked
clay tablets with wedge writing have been
found at the ruins of ancient cities of Sumer.

Teacher - Wedges - Axes to Airplanes

Integrate Sharp Science

Through centuries of trial-and-error testing, Japanese artisans learned how to make samurai swords without really understanding all the science involved.

Iron ore and charcoal(carbon) are heated (smelted). The furnaces can't get hot enough to fully melt the iron. Spongy blobs called "blooms" are cooled and broken into pieces. Iron without carbon is soft. Iron with too much carbon is brittle. Iron with just the right amount (0.15 to 1.5 percent) of carbon makes strong steel.

The samurai sword artisans do not understand the science but they do master the complicated, multi-step skills. They visibly sort out the different pieces for select purposes. Low carbon pieces are selected for the center of the sword that needs to absorb bangs and shocks (be tough and flexible). But this type of iron cannot be sharpened.

High carbon pieces are selected for the outside case. These pieces are strong & can be sharpened. But they are brittle so if the whole sword is made of this material, the sword will easily break. The iron bits are hammer forged together. This drives out impurities in the iron.

The soft center and strong outer case are forged together to make the sword. Thick clay is put on the back and faces but not the cutting edge.

The hot sword is quickly dunked (quenched) in water. The different parts of the sword cool at different times. This curves the sword. It is called "heat treating." The wedge-shaped cutting edge is sharpened on grinding stones by hand. To close, iron atoms mixed with carbon atoms are heated, repeatedly hammered, cooled quickly and then sharpened to make the sharp and resilient samurai swords.

Teacher - Wedges - Axes to Airplanes

2) Airplanes
—Teacher Guide—

With four forces, we fly. Our feet are no longer gravity-glued on the ground.

Indē Ed Project
Charitable Orgn

2) Airplanes
— Four Forces

Written by Douglas J. Alford
and Sally Kimangu
Illustrated by Cecil V. Bugayong

STEM-Zen Program

2) Airplanes
— Four Forces

With airplanes, our feet
are no longer gravity-glued
onto land. We earthlings
soar upwards to become
sky-lings.

1) 2) 3) 4)

Science of AIRPLANES

Easy Ideas 1

Airplanes 2

Cars 3

Computers 4

Smartphones 5

Food 6

Nature 7

Space 8

Light 9

AI 10

iii

STEM-Zen Program

Everyday Objects

Bonus

Airplane — 4 Forces

Airplanes
— Four Forces

When we fly, the forces are
with us: Forward, Backward,
Up and Down.
The breaths we take give
us a clue to the secret
ingredient of flight! — AIR!

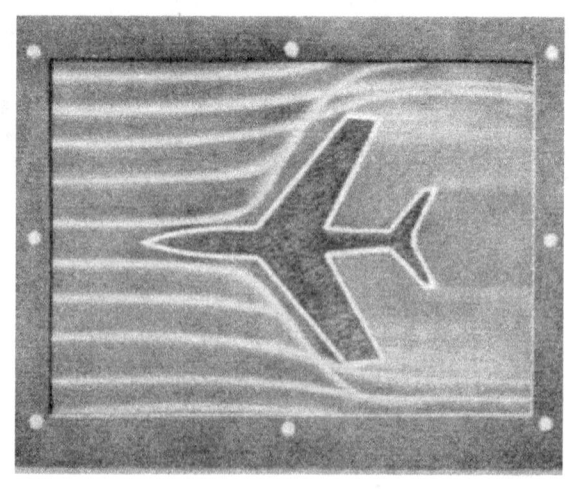

Airplanes — Four Forces

Table of Contents

Intro

This airplane weighs as much as 100 elephants. How can something so heavy get off the ground?

Why do airplanes fly?

Four Forces

It is all about
four uneven forces.

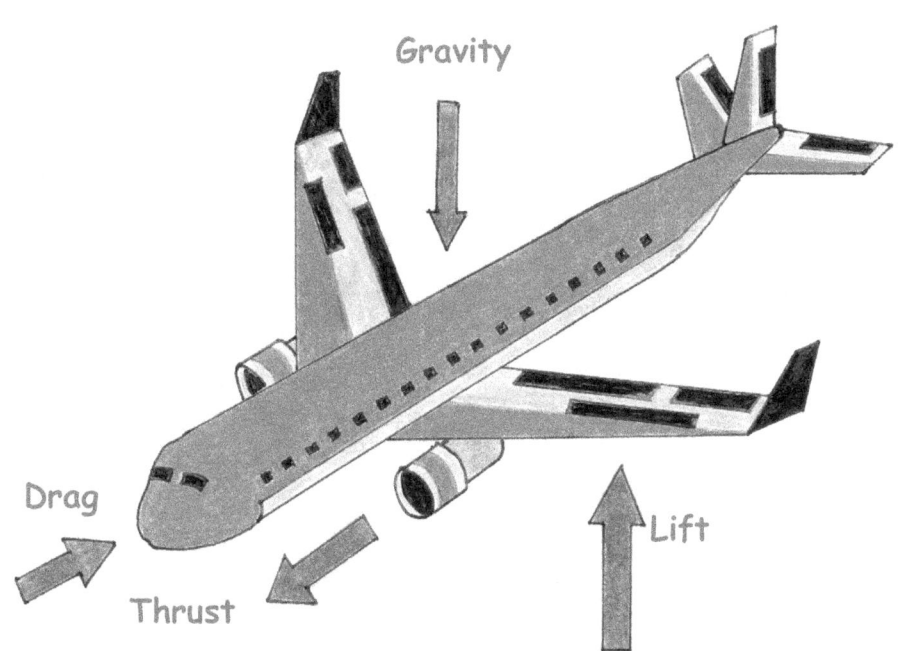

1) Thrust Forward
2) Lift Up
3) Drag Back
4) Gravity Down

1) Thrust Forward

Engines push the plane forward.
There are two types of engines.

Airplane — 4 Forces

Propellers

Prop planes have engines similar to cars. Burning fuel pushes pistons that turn propellers. The propeller is shaped like a twisted wing. Turning propellers, thrust the air backward. Thrust pushes the plane forward.

Jet Engines

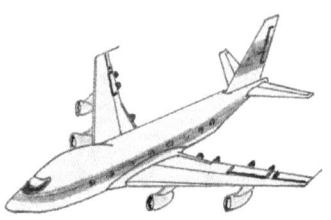

Air In   Squish Air   Burn   Exhaust

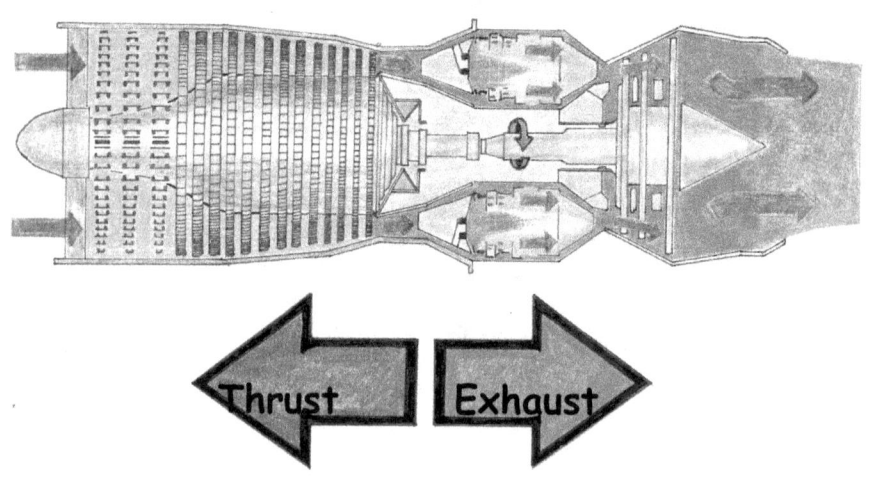

Thrust   Exhaust

Jet engines don't have propellers. Inside the jet engine, fans squish air into chambers. Next, the fuel and oxygen burn. The exhaust pushes backward. This thrusts the plane forward.

2) Lift Up

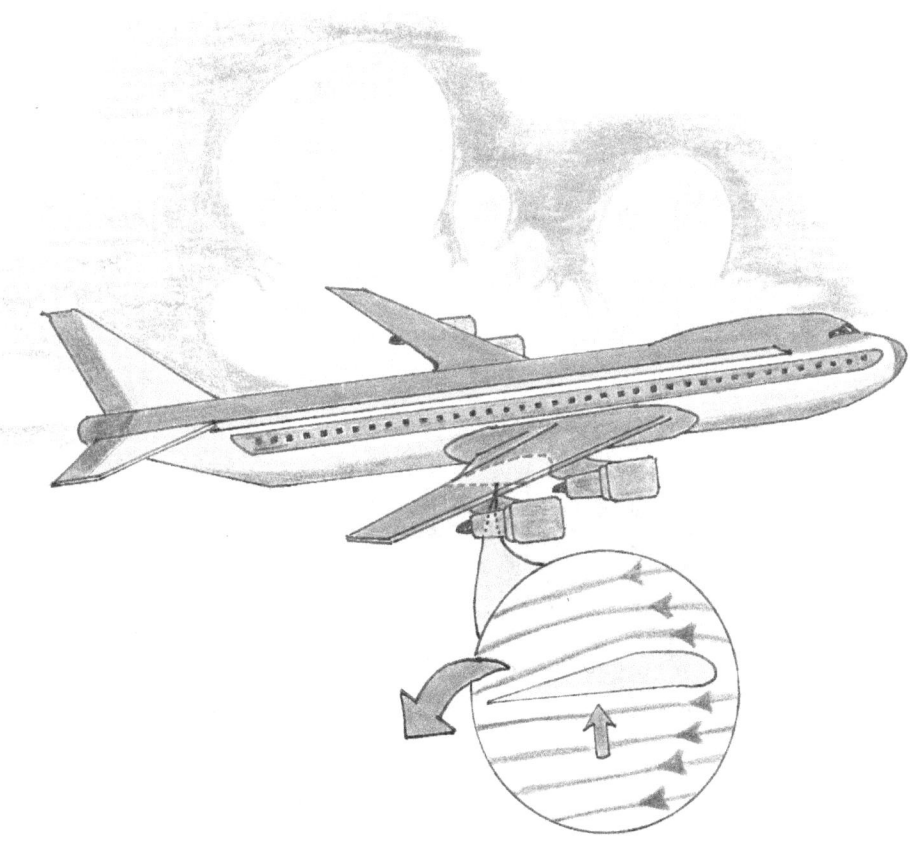

The shape of the wing forces air to flow faster over the top. There are two ways to think about how wings push or lift an airplane up. The faster air from on top flows over the wing and then down to push the plane up.

Wings Lift

Another way to see this, is that the faster air on top has lower pressure. The slower air under the wing is higher pressure. The higher-pressure air under the wing pushes the plane up. This is called lift.

Try this! Blow over a piece of paper. See how air pressure pushes or "lifts" the paper up.

Air Pressure

An airplane's first name is "air." Air is important to why planes fly. We live in a sea of air called the atmosphere. Earth's air is 60 miles (100 km) tall.

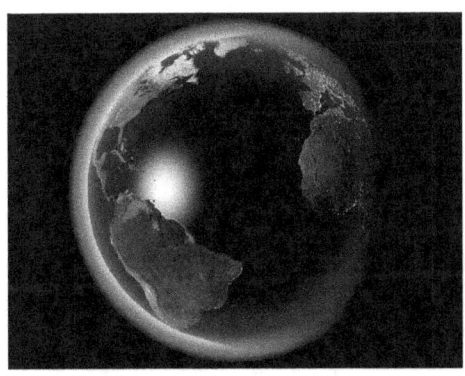

Air has pressure. To recap, differences in air pressure from above and below the wing push the plane up.

3) Drag Back

Air

Bike riders streamline their shape
so they go faster through the air.
Air in front of planes also pushes back.
This slows the plane and is called drag.
During flight, the engines have to
push harder to overcome drag.
Drag is useful to slow the plane
down when you want to land.

4) Gravity Down

A large object pulls on smaller objects. This is called gravity. Gravity pulls apples down from trees. Gravity also keeps us on Earth.

When a plane is flying, slowing the plane leads to less lift. Gravity then pulls the plane down.

Uneven Forces

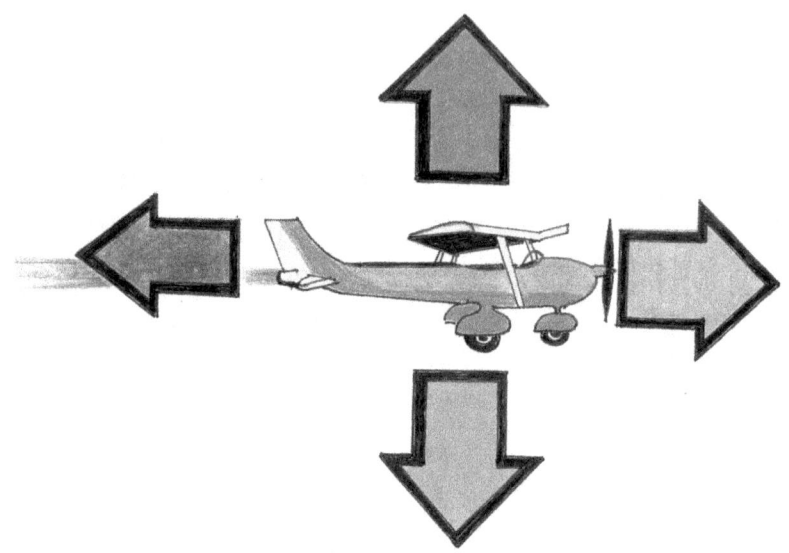

These are the four uneven forces of flight. Let's look at how they make a plane — with the weight of a herd of elephants — fly.

Taxi

From the gate, the plane is slowly
pushed to the runway. The engine
power increases. Thrust pushes
the plane forward faster and faster.

Weight Off Wheels – WOW!

Air flows fast over the wings.
Lift pushes up as gravity pulls down.
There is a WOW point, where lift up is
greater than gravity down. Suddenly, the
plane lifts off the land. The Weight is
Off the Wheels. WOW! The plane flies.

Take Off

Next, lift continues to push
the plane up into the air.

Cruise

There is a point where lift up and gravity down are even or balanced. The plane stops climbing. Engine thrust continues to push the plane forward.

Lift up equals gravity down. Forward thrust is more than the drag back.

Airplane — 4 Forces

Control

During flight, controls
change the plane's direction.

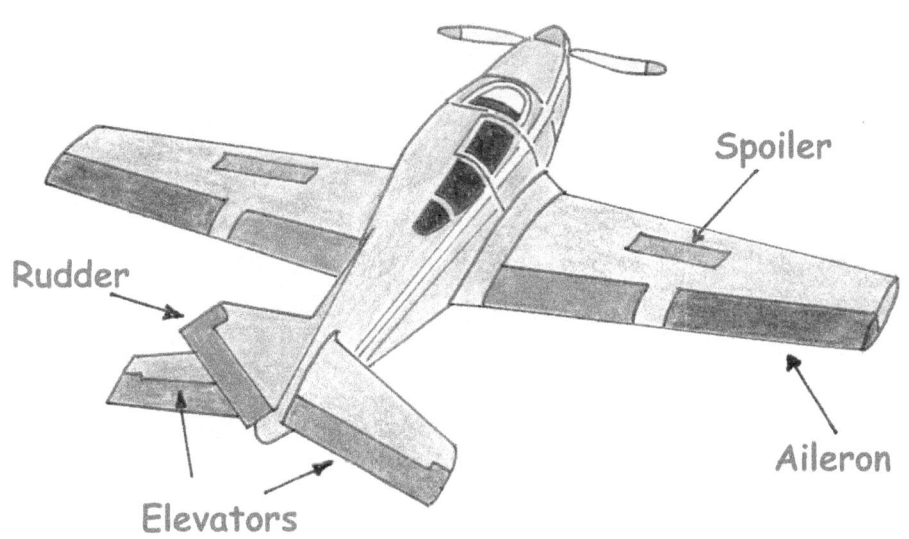

Rudder

Spoiler

Elevators

Aileron

Rudder <u>YAW</u>

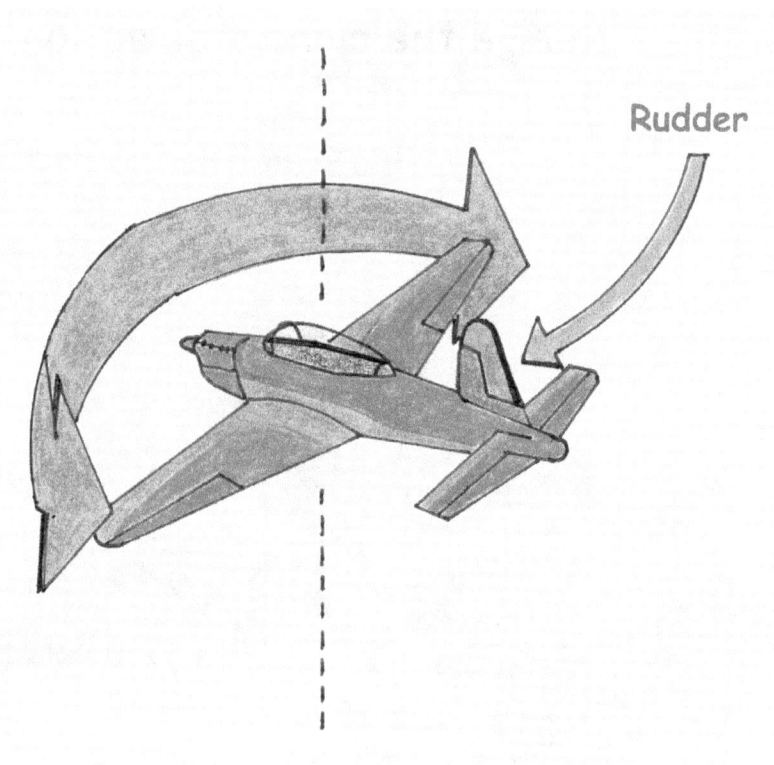

Rudder

Rudder moves the
plane left or right.

Elevator <u>PITCH</u>

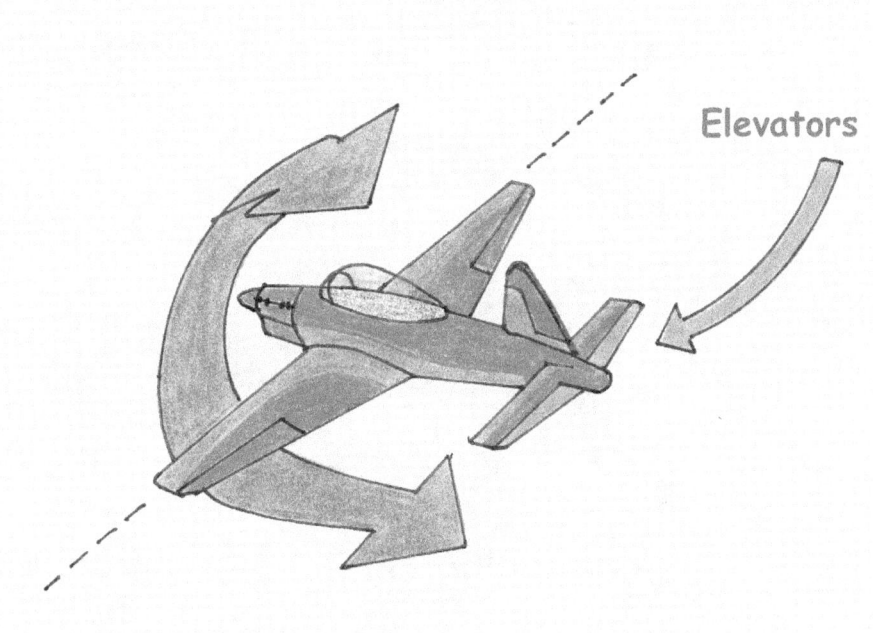

Elevators

Tail elevators move
the plane up or down.

Aileron <u>ROLL</u>

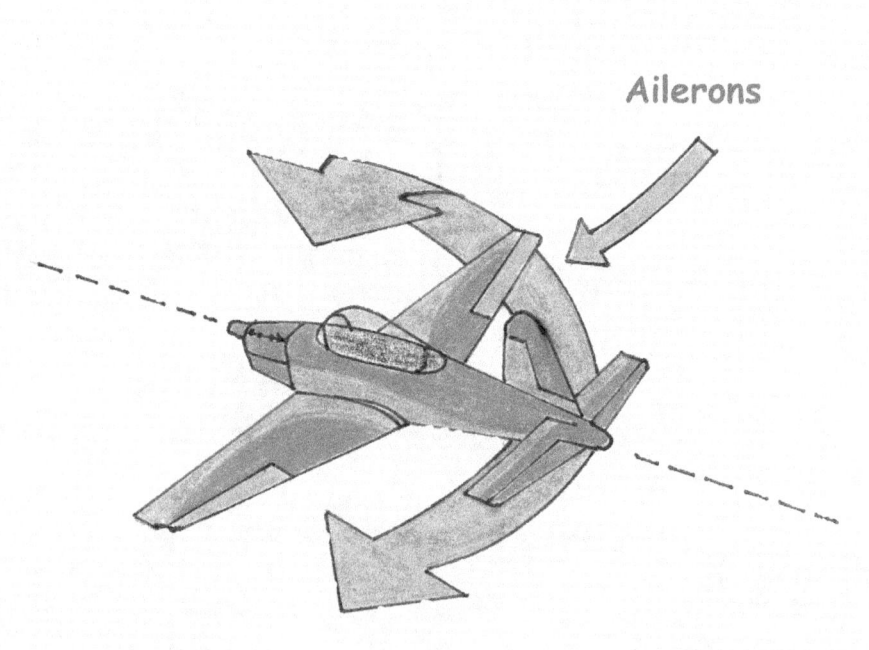

Ailerons

Ailerons roll the
plane from side to side.

Combo Controls

PITCH

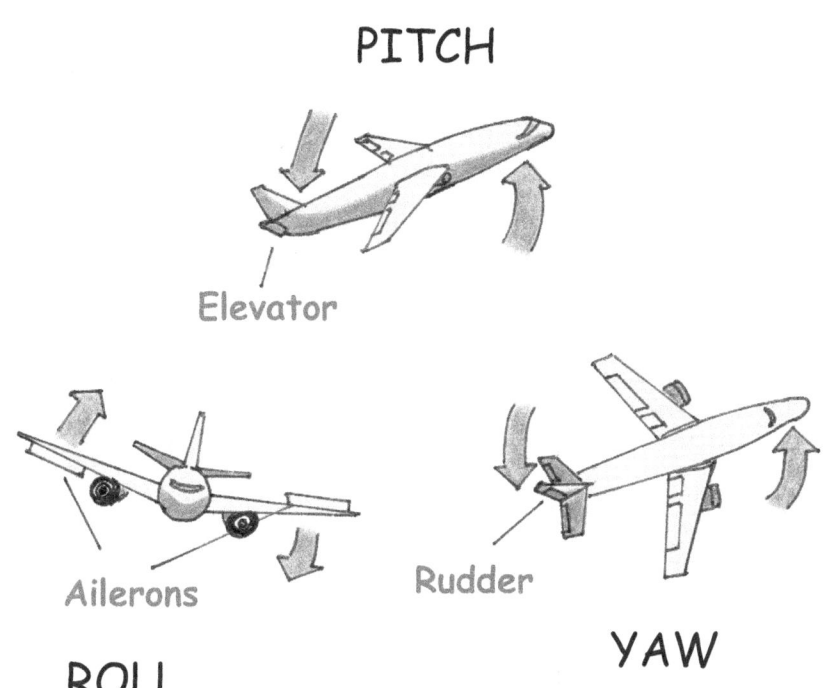

Elevator

Ailerons

Rudder

YAW

ROLL

Pilots combine the three flight controls to change the plane's direction smoothly. This is more comfortable for the passengers and crew.

To recap, pilots call the three control directions: Yaw, Pitch and Roll.

Navigate

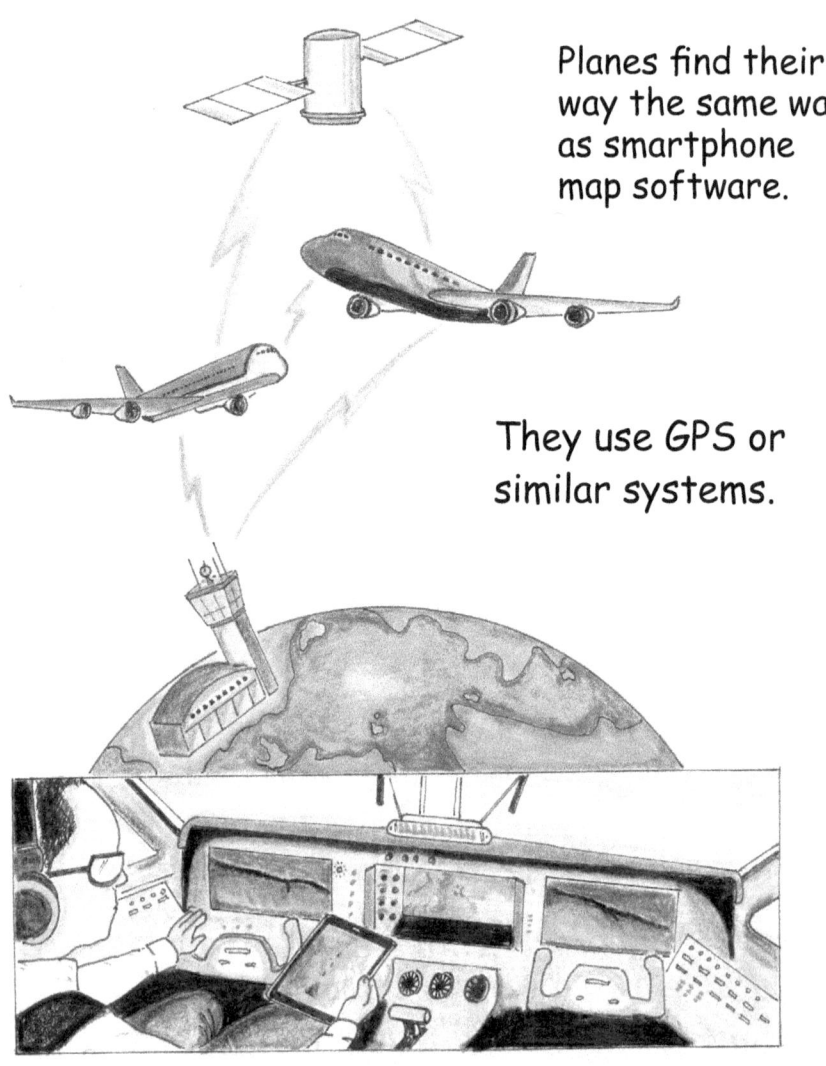

Planes find their way the same way as smartphone map software.

They use GPS or similar systems.

GPS or Global Positioning System is a ring of satellites that tell location on and around the Earth. This satellite navigation is called "sat-nav."

Communicate

Pilots in airplanes communicate by radios with ground stations. They discuss the plane's flight status. Pilots also ask for permission to land.

Gravity Down

So how do you bring the plane back to land?

Slowing the engines, reduces thrust and reduces air flowing over the wings which reduces lift.

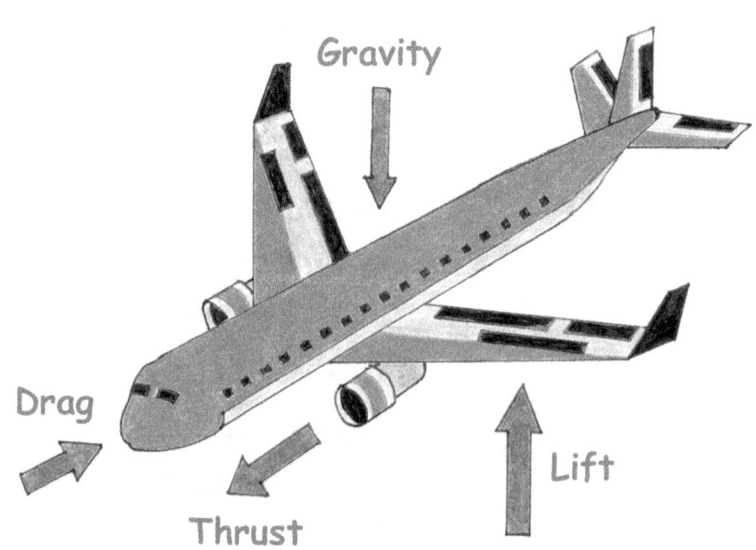

Gravity is now stronger than lift.
Gravity pulls the plane down.

Airplane — 4 Forces

Drag Back

Spoiler flaps increase drag.
This reduces the plane's speed too.

spoilers

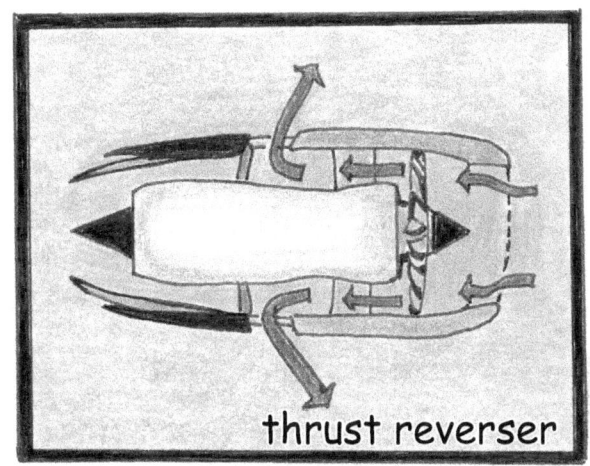

thrust reverser

On the ground, engine thrust reversers
and brakes help slow down the plane.

Steer and Stop

Like a car, steering moves the
plane right or left on land. Brakes
stop the plane at the new airport gate.

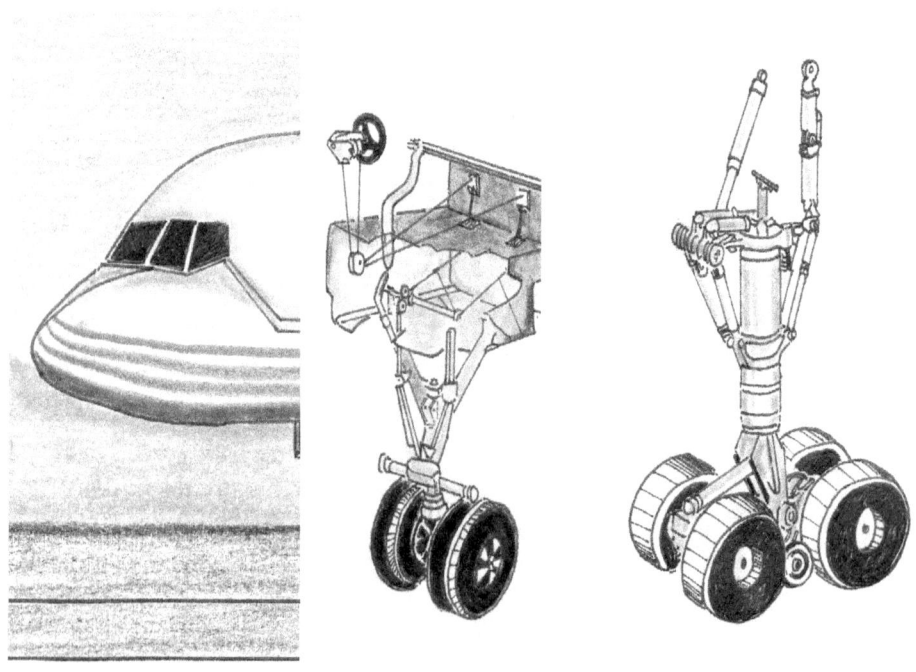

And so the story of flight has
come full circle, from the land
to the sky and back to land again.

Forces of Flight

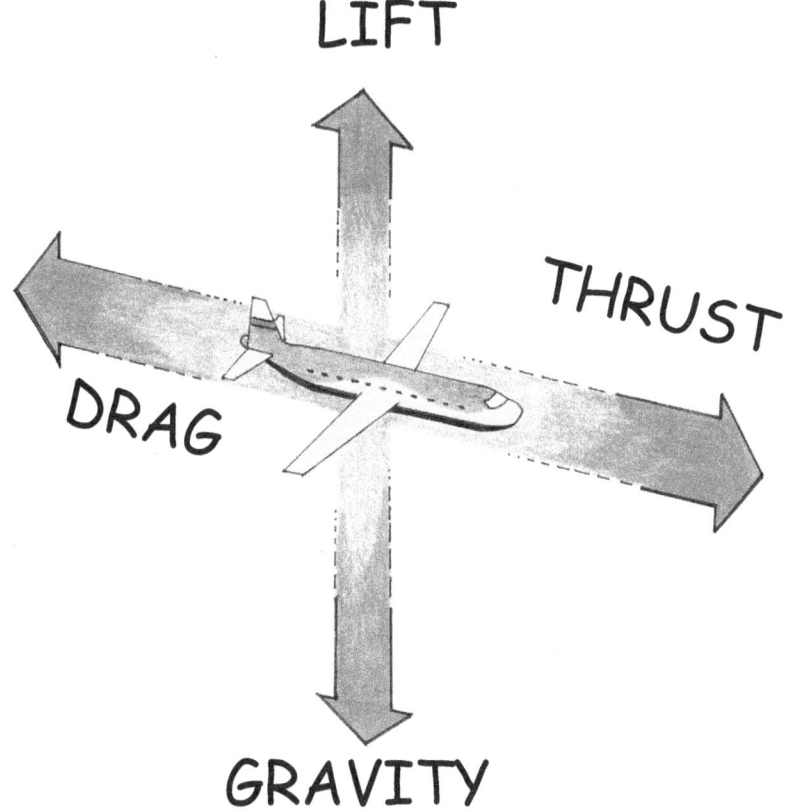

To recap, the secret
science of flight is simple.

Engines thrust forward.
Wings lift up.
Drag pushes back.
Gravity pulls the plane down.

Conclusion

The first ingredient - of why planes fly -
is air. Engines force air backward to
push the plane forward. Air flows over and
under the wings to lift the plane up. Air
resistance drags the plane to slow it down.

Gravity brings the plane, back down to
Earth. Gravity also holds a circle sphere
of air around the Earth. Planes fly
around the world in this atmosphere.

Now you know why heavy airplanes fly!
Simply said, flight is the right combination
of air pressure, thrust, lift, drag and gravity.

Here are eight interesting
questions about flight.

1) In four words,
why do airplanes fly?

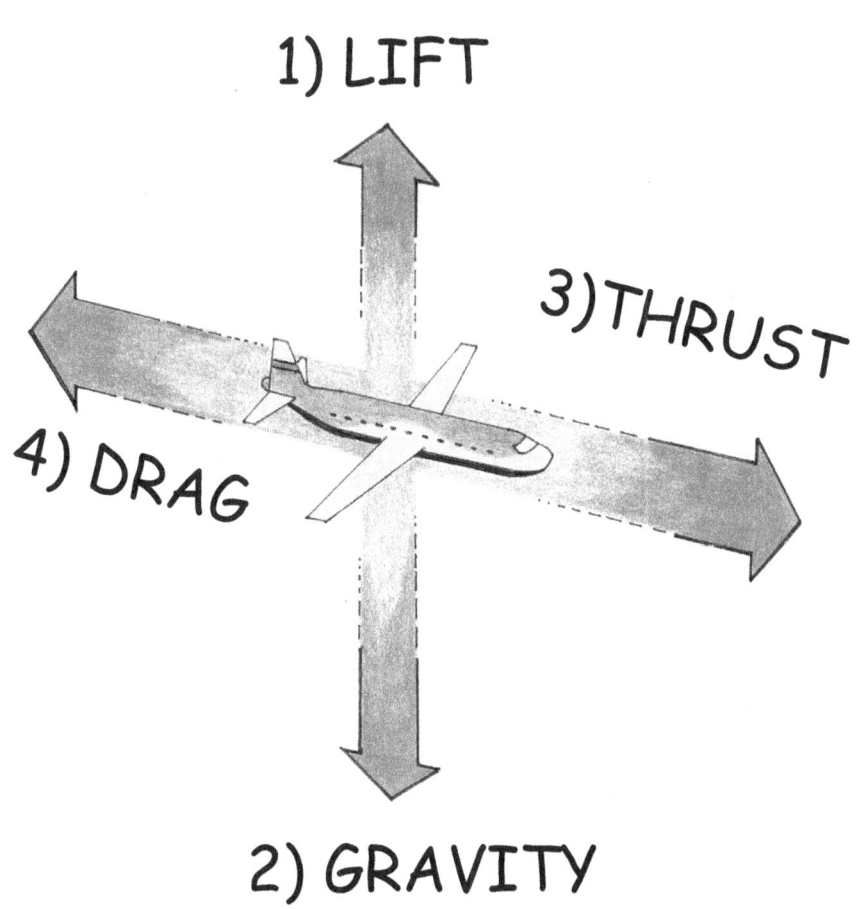

1) LIFT

3) THRUST

4) DRAG

2) GRAVITY

2) What are three flight control directions called?

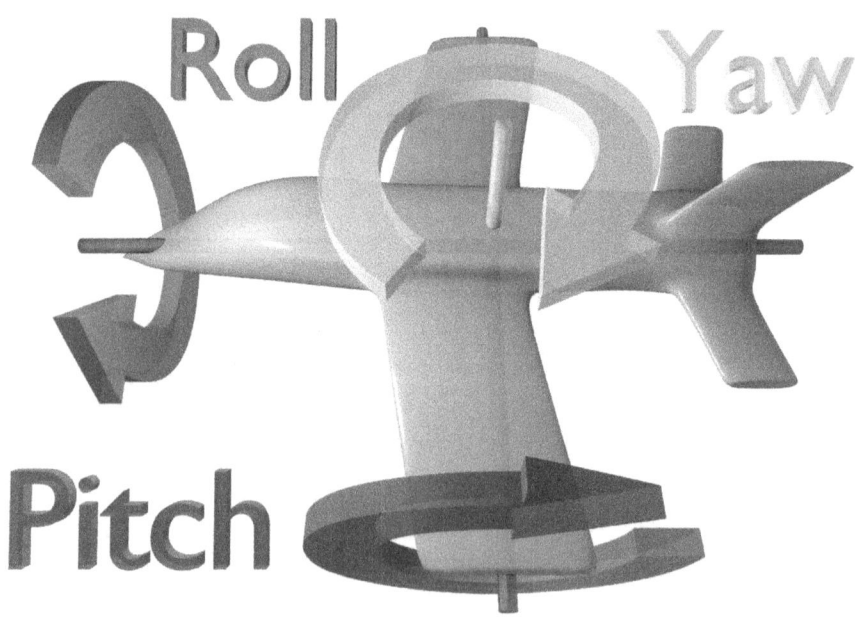

3) What are the three
parts that control the plane?

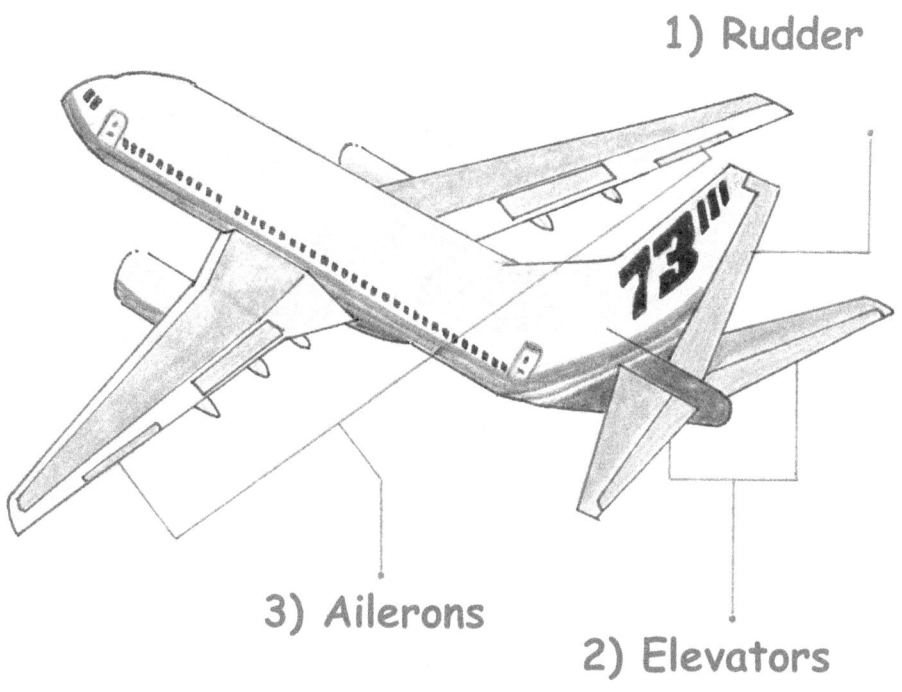

1) Rudder

3) Ailerons

2) Elevators

4) Why do jet airplanes have swept wings and engines hanging below?

Propeller planes have straight wings but jet wings are not straight. Jet engines are so strong they can cause a problem called flutter. Flutter is uncontrolled bending up and down of the wing. This can break the plane. Swept wings and engines below the wing, fix this problem.

5) Why did the first commercial jet aircraft called "Comets" crash?

Metal fatigue caused cracks in the fuselage at the corners of the <u>square</u> windows. This caused the planes to crash. Today, planes have **<u>oval</u>** windows.

6) How much does a
wide-body commercial
aircraft weigh?

x 100

At take-off, a fully loaded wide-body airplane
weighs as much as one hundred elephants.

7) Why does the wing`s shape change during flight?

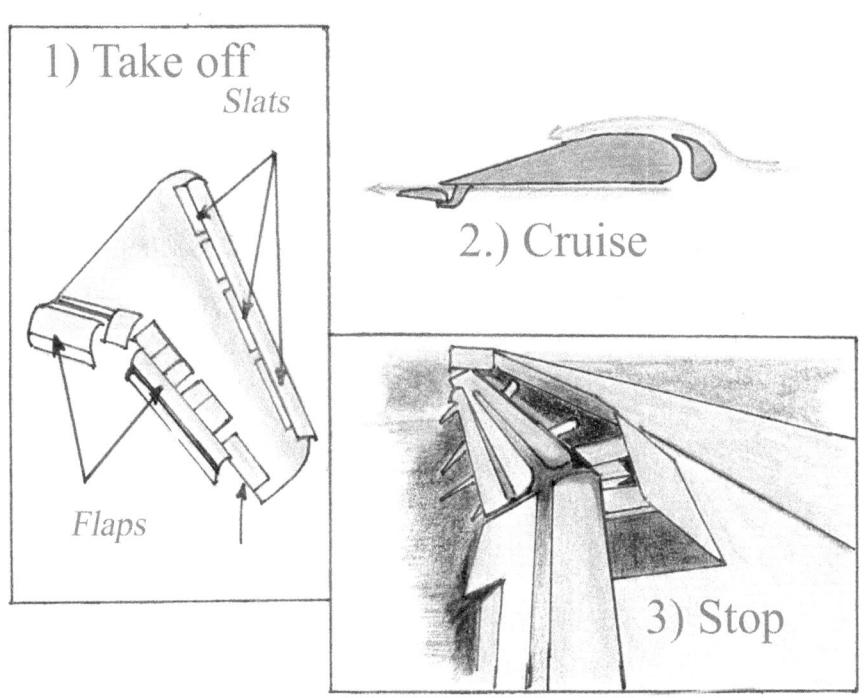

1) Take off
Slats

Flaps

2.) Cruise

3) Stop

The wing changes shape to:
1) increase lift during take off;
2) balance forces while cruising; and
3) slow down during landing.

8) Why do commercial planes fly below the speed of sound?

The speed of sound is about 1,234 kilometers per hour (767 miles per hour). At this speed, drag increases and the airplane is harder to control. Commercial planes are not designed to go at or faster than the speed of sound. This picture is a fighter plane going faster than the speed of sound. It is also called breaking the sound barrier with a sonic boom!

Airplanes

Credits

Pages Description

1,33   (N6067E)_wide_angle_by Dave Subelack

Front Cover, 2-7,9-11,15-26,28,30,31,33 Elephant, 34,

Back Cover illustrated by **Cecil V. Bugayong**

29 Wikipedia by ZeroOne

36 A380 by Maarten Visser

All other pictures are in the **Public Domain**.

Airplanes – ONE PAGER

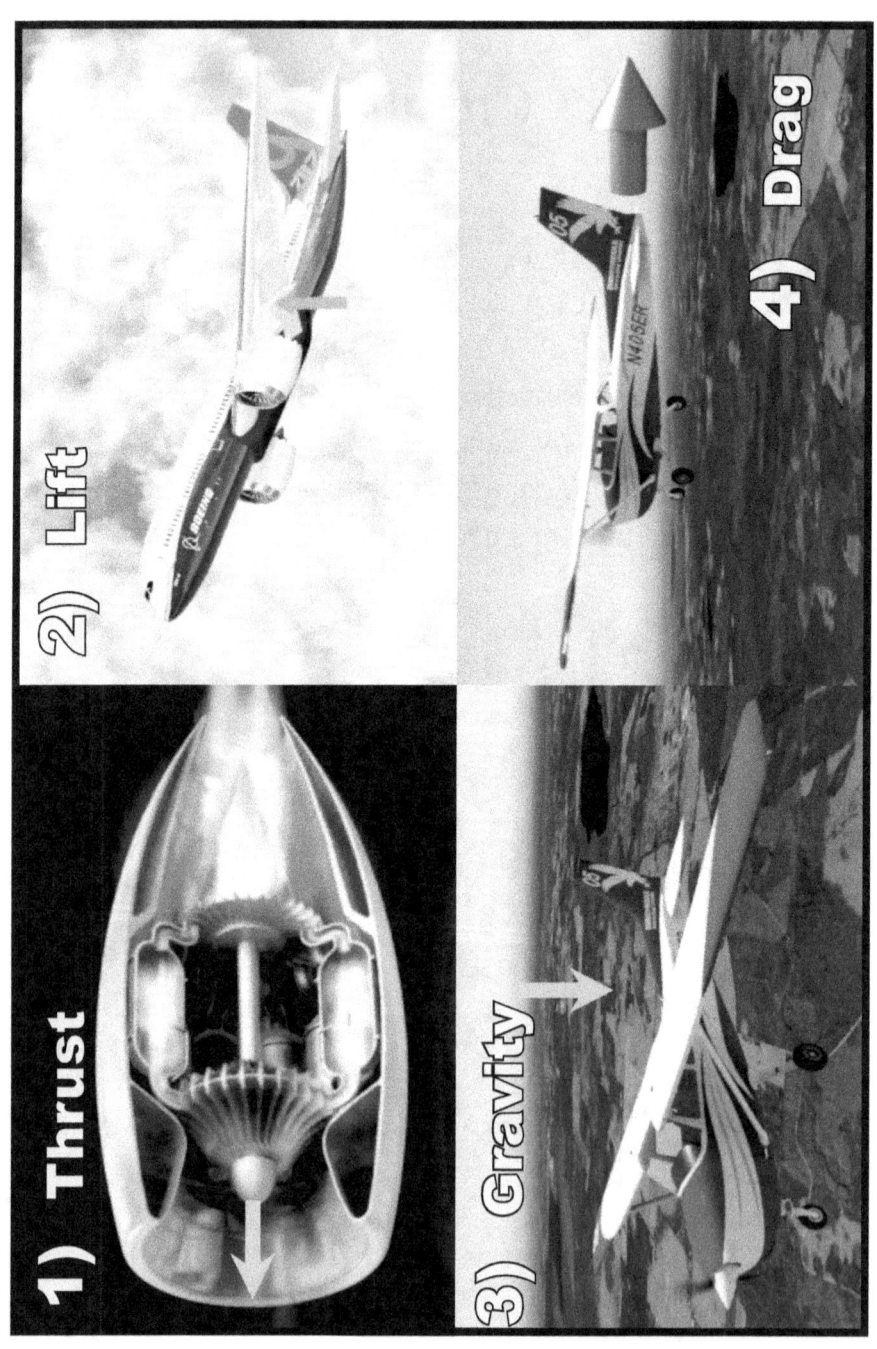

Watch VIDEO

For a copy of this video contact:

2) Airplanes — Four Forces

Somehow seeing birds fly makes sense to us. But what about an airplane that weighs as much as an elephant herd? How can heavy planes fly? See inside this book for the science of: engine push, wing lift, drag back and gravity down. Moving air is the secret ingredient that enables flight.

Science

EDUSTORE AFRICA

Indë Ed Project
Charitable Org'n

2) Airplanes
— Past and Present

People think about engines and energizing air.
Next, our planes soar into the sky!

Douglas J. Alford
Sally Kimangu

STEM-Zen Program

Airplanes
From Bird Flaps to Human Flights

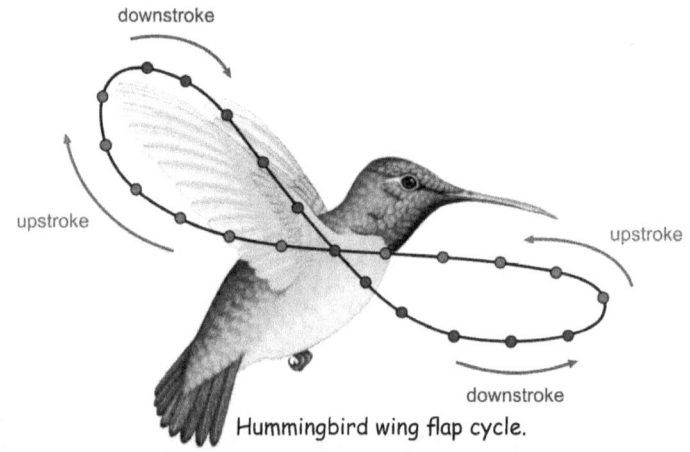

downstroke

upstroke

upstroke

downstroke

Hummingbird wing flap cycle.

From Flaps to Flights

For ages, we earthlings stare in awe at the birds who
fly above us. Many people tried but failed to fly.
Today, planes fly us around the world in
hours. Our feet are no longer gravity-glued
to the ground. We are sky-lings now!

Present tense verbs are given priority in
this book to simplify international versions.

Planes — Past and Present

Table of Contents

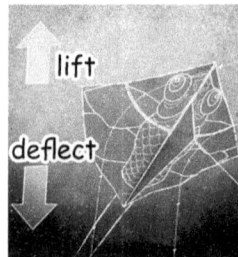

lift

deflect

Air deflects down
to lift a kite up.

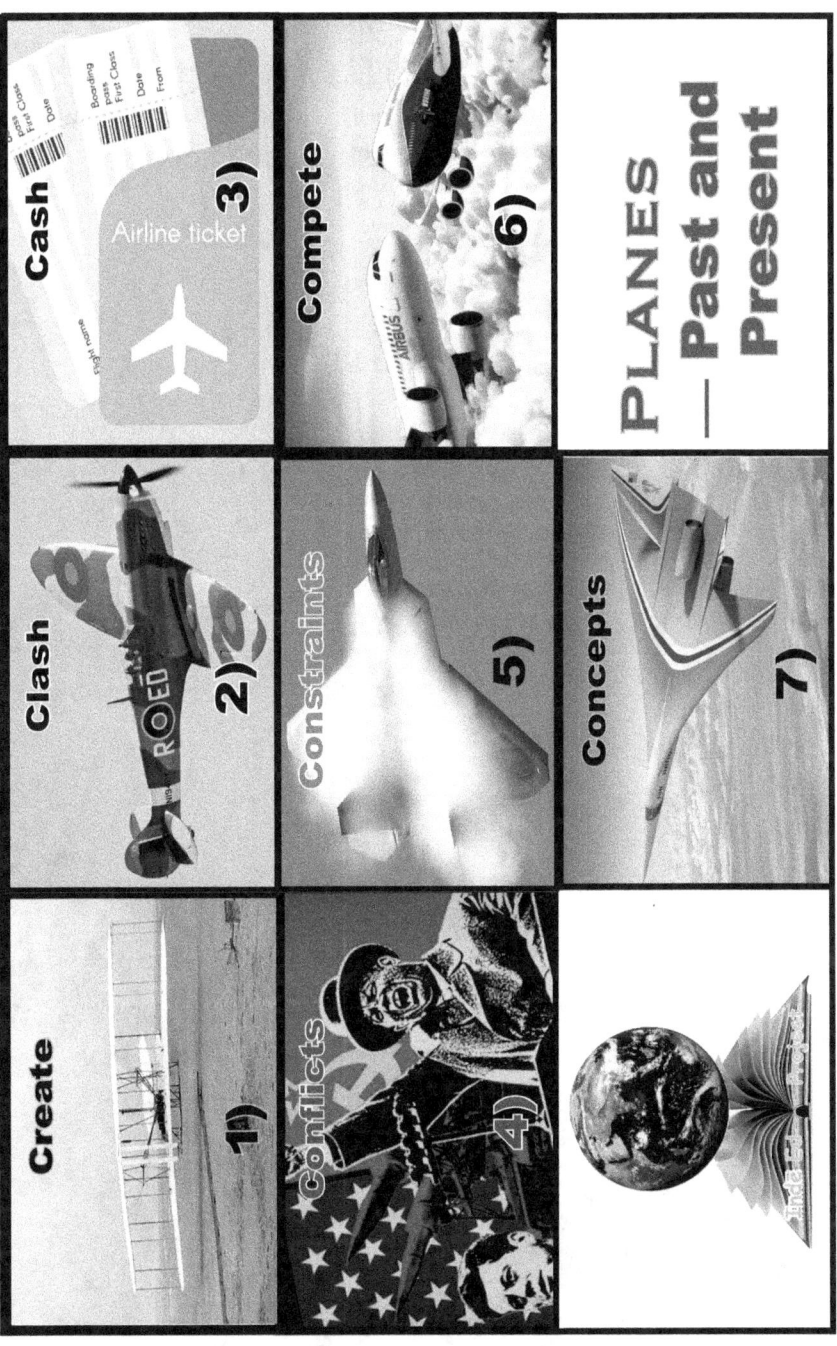

PLANES
— Past and
Present

Create
1)

Clash
2)

Cash
3)

Conflicts
4)

Constraints
5)

Compete
6)

Concepts
7)

Airline ticket

Feather Flaps

Flight is all about air pressure. Birds flap their feathers to push air down. The way air flows over and under wings pushes birds up.

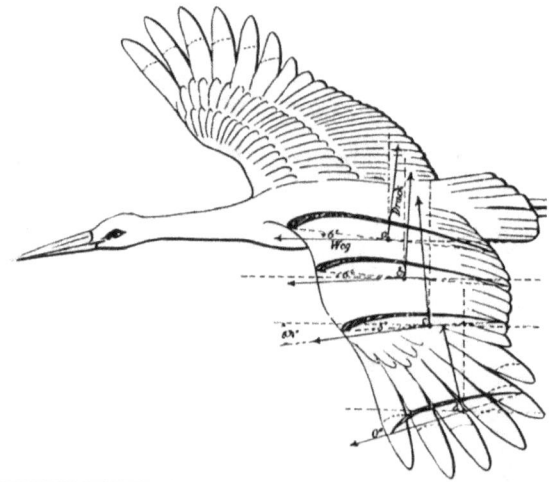

Flapping feathers thrust birds forward too.

Someone noticed gulls gliding at the seashore.

Cayley's Craft

In England, Sir George Cayley studies birds and how they fly. In 1853, 80-year-old Cayley makes a glider that can carry a person. The story is told how he has his coach driver "fly" in the glider shown below.

artist concept

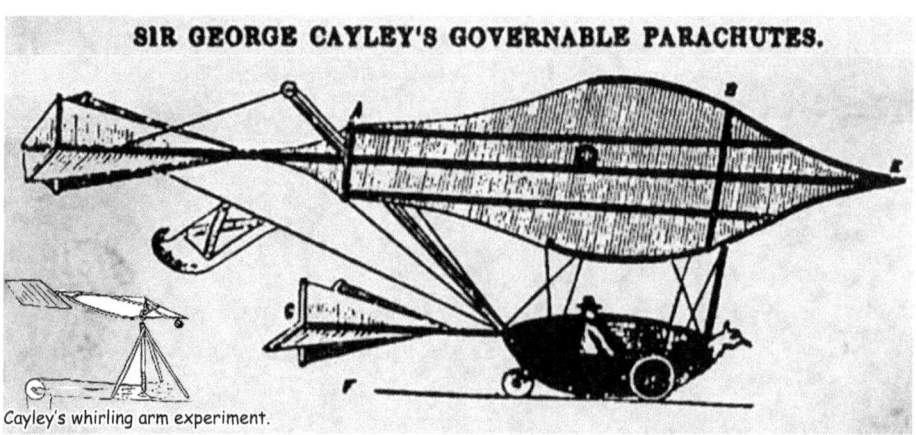

SIR GEORGE CAYLEY'S GOVERNABLE PARACHUTES.

Cayley's whirling arm experiment.

After a brief glide, the glider crashes. Next, the shaken but unhurt
chauffeur quits his job. He said he is hired to drive the horse coach,
not to fly. In another version of the story it is a local boy who glides first.
Either way it is the first recorded flight of a glider carrying a person.

Toy Gliders

A father gives his young sons a rubber band-powered toy glider called a planophore. Orville and Wilbur Wright are fascinated with flight from this point on!

planophore

Fair Use. PBS

Penaud

ONE) Create
Wright Gliders

As adults, the Wright Brothers make their own life-sized gliders. They start with Cayley's design. Then they create their own. They build wind tunnels to test their ideas. They pay for their own trial-and-error flying experiments. Crashes are frequent.

Wright Flight

In 1903, they go to Kitty Hawk. The
wind at the beach gives them extra lift.
The sand gives them a safe place to land.
To save weight, their airplane doesn't
have any tires. The Wright Brothers
are the first to fly an airplane.

Propellers, Pistons & Push

The Wright Brothers use an internal combustion engine. It is made in their machine shop that is part of their bike business. The pistons move the propellers that push the plane forward.

Different air pressures push the plane up.

The Wright Brothers create their propeller shape by twisting the cross-section of the wing.
Propellers push the plane forwards.

Propellers push the plane forwards.

Court & Competition

The Wright Brothers fly their planes around the USA and Europe. Other companies start making airplanes. Wilbur sues their USA competition to protect the Wright Brothers patents. While the lawyers argue in court, the USA falls behind in the airplane race.

Notice how the first airplanes do not have interiors.
Everything and everyone is outside and open to the weather.

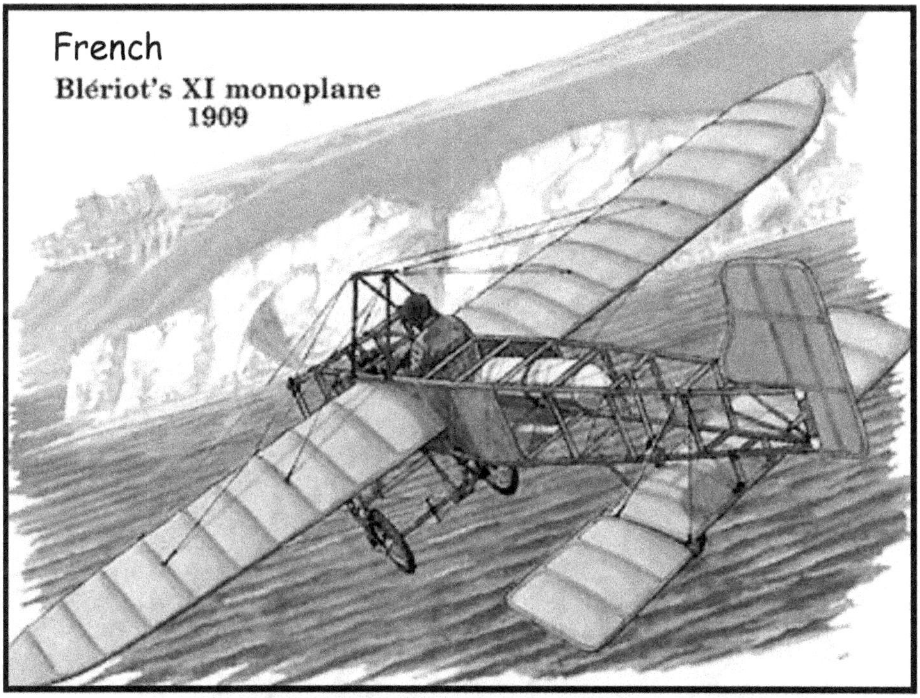

French
Blériot's XI monoplane
1909

WWI — French Flies

When the USA enters World War I, they have no war planes of their own. The USA has to buy planes from France and England.

Airmail

After World War I, the US Post
Office uses private planes to transport mail.
They call the quicker service "airmail."
Like the Pony Express of the 1860s, airmail
in the 1920s is very dangerous. Many
planes crash and the pilots are killed.

Orteig Prize

In the 1920s, Raymond Orteig is a rich
New York hotel owner. He offers a prize
of $25,000 to the first person who can
fly non-stop from New York City to
Paris or vice-versa. The prize interests
an airmail pilot named Charles Lindbergh.

Lucky Lindy

Six pilots die trying for the
Orteig Prize. Charles Lindberg,
known as "Lucky Lindy," wants to try.

He talks St. Louis businessmen
into investing in his flying adventure.

Lindy works with the Ryan Company to
build his plane for $15,000. The official
name of the plane is "Ryan NYP" which
means New York to Paris. Lindy calls
his plane, "The Spirit of St. Louis."

In 1927, Lindy takes off from New York City with 450 gallons of gas. For food, he takes a few sandwiches with him. He said he will get more food in Paris or if he doesn't make it to Paris, then he will not need more food because he will be dead. He flies alone for the next 33 hours. He flies 3,600 miles (5,800 kilometers).

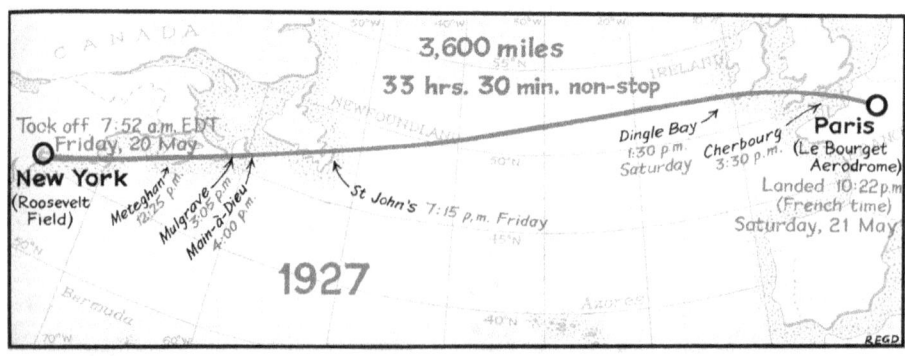

Smithsonian National Air & Space Museum

He lands in Paris and is cheered
by a crowd of over 150,000 people.
Lucky Lindy is a worldwide hero.

Wiley's World Record

After Lindy's success, more people start breaking other flight records.

In 1933, Wiley Post flies by himself around the world in less than 8 days.

As a side note, in 1926, Wiley loses his eye in an oil rig accident. He uses the lawsuit settlement money to buy his first airplane.

After his round-the-world solo trip, Wiley is
greeted in New York City by 50,000 waving people.

1933

deep-sea diving suit.

In 1934, Wiley wears a pressurized, deep-sea diving suit to fly his unpressurized airplane at high altitudes.

TWO) Clash
World War II

In the 1940s, many new planes are created for the global clash of countries called World War II.

Blitzkrieg

The Nazis use fighters and
ground attack-aircraft in their
lightening warfare called "blitzkrieg."

Pearl Harbor

The imperial Japanese use fighter, bomber and
torpedo planes to attack the USA at Pearl Harbor Hawaii.

90 years before Pearl Harbor, Japan is a closed country with few factories and little technology.

Planes — Past and Present

Fighters

World War II fighter planes
fight each other in the air.

The British have a secret weapon during
their fight with the Nazi Air Force.

The British invent RADAR that uses
radio waves to detect enemy planes.
The British win the Battle of Britain even
though the Nazis have three times as many planes.

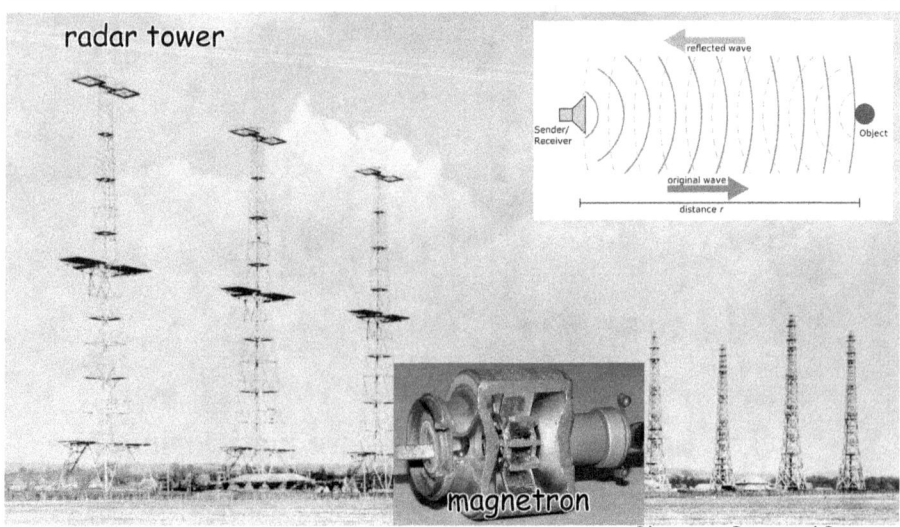

radar tower

reflected wave

Sender/
Receiver

Object

original wave

distance r

magnetron

Planes — Past and Present

Flying Tigers

In WW II, American volunteers called the "Flying Tigers" use their shark-painted planes to help protect and defend China.

Himalayan Mountains called "the hump."

The Allies fly supplies to China over the Himalayas from India.

Bombers

During the war, both sides make bomber airplanes. They are effective at the job they are designed for. Air power is critical to winning battles in World War II.

Most WW II planes have unpressurized interiors which means they are open to the outside air. Unpressurized planes can only fly up to an altitude of 10,000 feet (3,000 meters). Remember that air gets thinner – has less pressure – the higher up you go in the atmosphere. The advantage of high altitude flying is smoother flight, less drag and less fuel burned.

Pressurized Insides

Both Allies and Nazis each make their own pressurized bombers that can fly at high altitudes. Nazis only build a few of their planes. The USA makes thousands of B-29s. In 1945, B-29s drop atomic bombs that end World War II in Asia.

Atomic Bomb

THREE) Cash
Passenger Planes

After WW II, private companies, motivated by cash, make planes to fly the paying public. These planes with pressurized cabins fly higher than storm clouds, so the ride is smoother.

Notice the straight wings and propeller engines.

First Jets

The roots of faster planes go back to World War II. Based on British engine designs, Nazis make the first operational jet fighter planes. A quantity of only a few hundred see combat before the end of the war. The Nazi jets have high performance. But the Nazis make too few of these planes to impact the war outcome.

Messerschmitt Me 262

Through wind tunnel testing, the Nazis learn that swept wings prevent flutter problems with jet engines. Flutter is uncontrolled bending up and down of the wing. This can break the plane. Swept wings with engines below fix this problem.

FOUR) Conflicts
Cold War

After World War II, a competition or Cold War breaks out between the Western Allies and Eastern Communist countries. As part of the Cold War, the race is on for better airplanes. This includes new fighters and bombers.

Uncle Wiggly Wings

After World War II, Germany is divided into the democratic west and communist east. The capital city, Berlin, was also divided. It is located in the communist eastern part of Germany. In a move to get control of the whole city, the communists cut off supplies to the western parts of the city. The war-torn city is in ruins and the people are starving. The West starts flying supplies into Berlin by air. A young American named Gail Halvorsen is one of the pilots bringing food into Berlin. He sees children at the airport fence. He gets an idea.

1948 - 49

32 USA

Berlin Airlift delivers food and fuel in 1948-49 blockade

Templehof Airbase during the Berlin Airlift.

Candy Bomber

The next day, he drops candy attached to
handkerchief parachutes from his plane to the
children below.   Later, he speaks with some
of the children. He promises to drop more
candy from his plane the next day. The
children ask, "There are hundreds of planes flying
into Berlin, how will we know which one is yours?"
He replies, "I will wiggle my plane's wings." The
children call Pilot Halvorsen, "Uncle Wiggly Wings."
Other pilots soon start dropping candy too. The
story makes news in the USA. Many people donate
candy. The Berlin Airlift lasts about one year.
During that time, pilots drop over 23 tons
of free chocolate and candy to the children.
More than just sweets, the candy gives hope
to those who receive it and those who share.

These pictures show "Uncle Wiggly Wings" dropping handkerchief parachutes of candy to children in Berlin during the Cold War.

How Jets Work

Up to this point, most planes use piston engines. Piston engines are very complicated.

In the 1930s, a British inventor creates a jet engine. It takes years to perfect and put into production. Jet engines are more simple and powerful.

Fans compress air that ignites in a chamber. Hot exhaust flows out the back and pushes the plane forward.

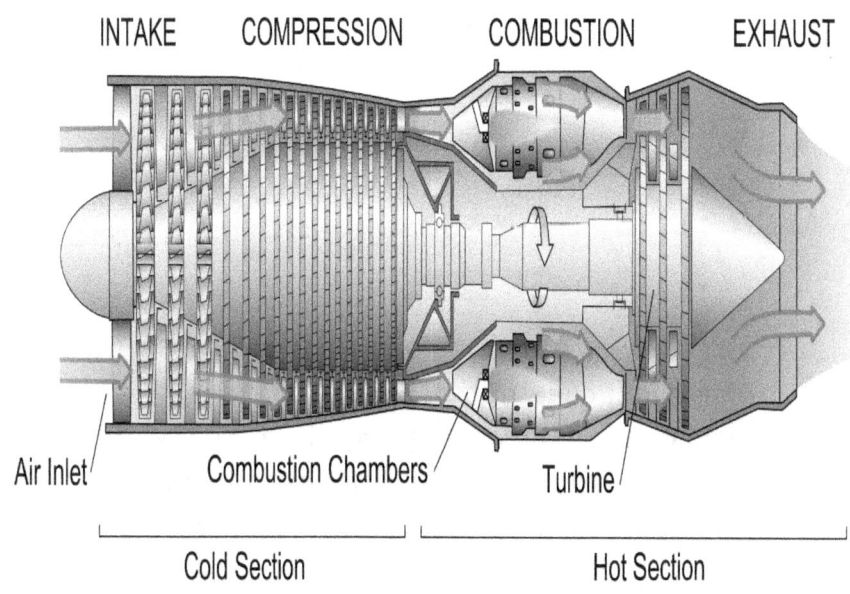

INTAKE   COMPRESSION   COMBUSTION   EXHAUST

Air Inlet   Combustion Chambers   Turbine

Cold Section   Hot Section

Flutter

Propeller planes have straight wings. Jet engines are more powerful than propeller engines. Most jets have wedge-shaped swept-back wings.

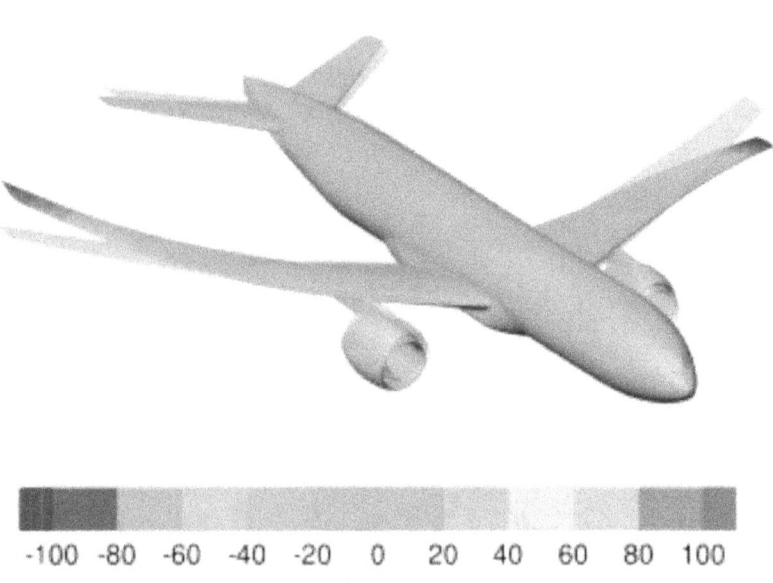

-100 -80 -60 -40 -20 0 20 40 60 80 100

Jet engines are so strong they can cause a problem called flutter. Flutter is uncontrolled bending up and down of the wing. This can break the plane. Swept wings with engines below fix this problem.

Comet Cracks

The first passenger jets are called "Comets." They have square windows. After thousands of flights, the square windows lead to stress cracks in the corners. Airplanes crash. People lose confidence in the Comet.

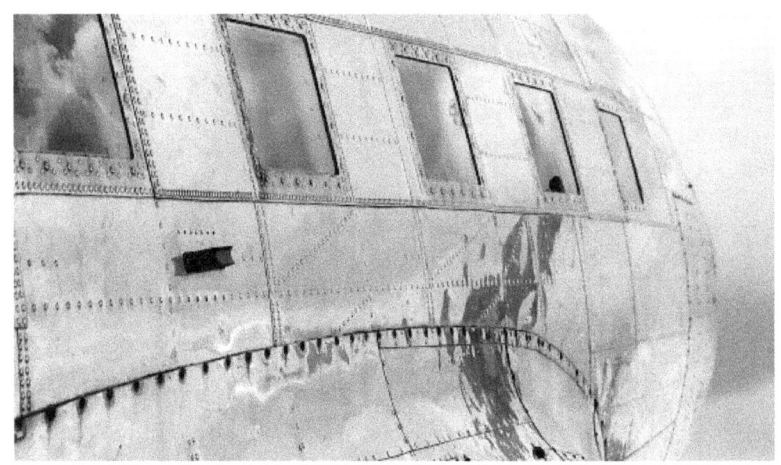

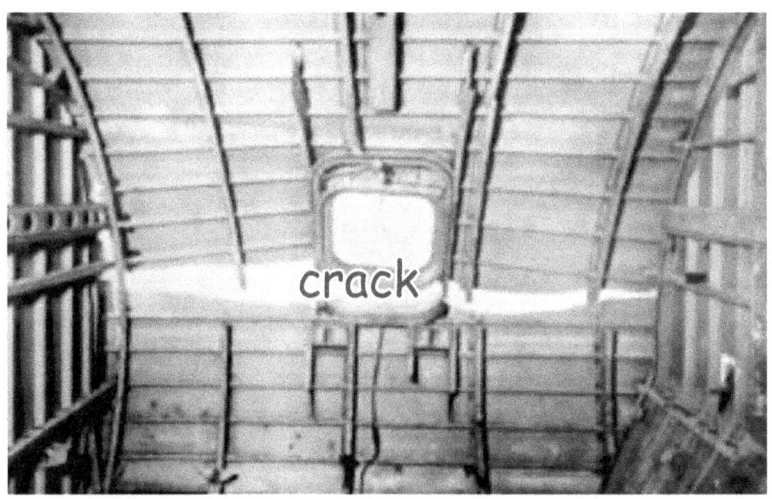

crack

Passenger planes are redesigned with oval-shaped
windows which don't get stress fractures.

707 Jet Plane

Based on the military tanker KC-135,
Boeing makes the 707 passenger planes.
The 707 is the first successful jet
airliner. It has oval windows.

KC-135

707

Why do most passenger
planes go slower than 700 miles
(1,130 kilometers) per hour?

Wikipedia by Dylan Ashe

FIVE) Constraints
Sound Barrier

Sound travels at a speed of 768 miles (1,236 kilometers) per hour depending on altitude.
Many people believe that an airplane can <u>not</u> go faster than this. Many guess that the plane will break apart or at least crash at these speeds. NASA wants to try to break the sound barrier.

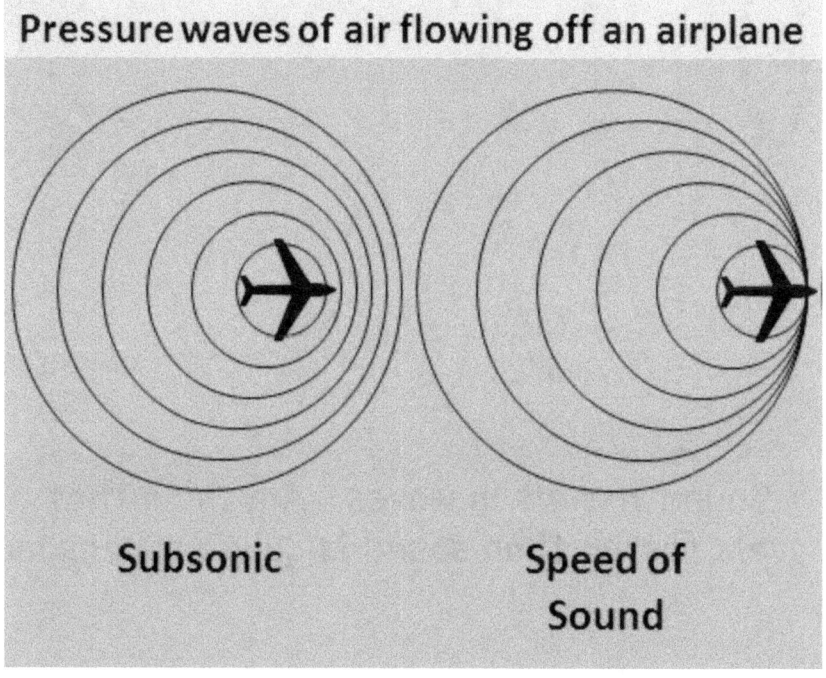

Pressure waves of air flowing off an airplane

Subsonic   Speed of Sound

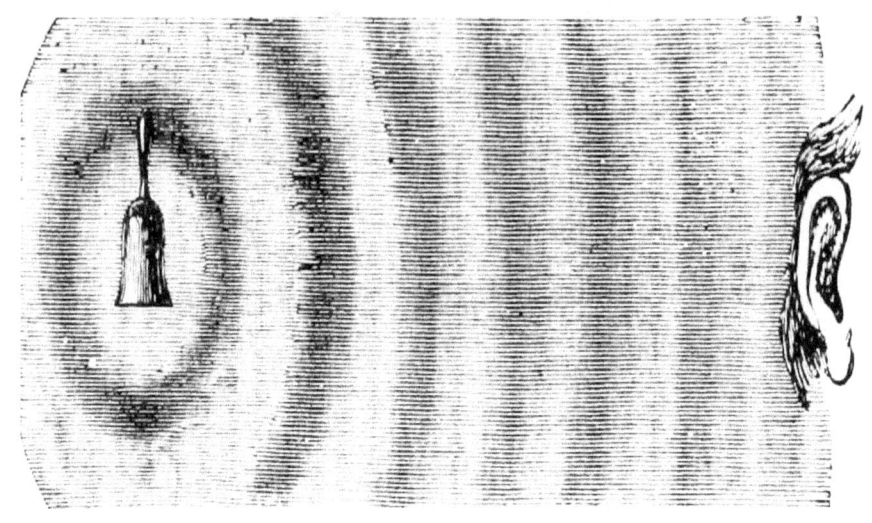

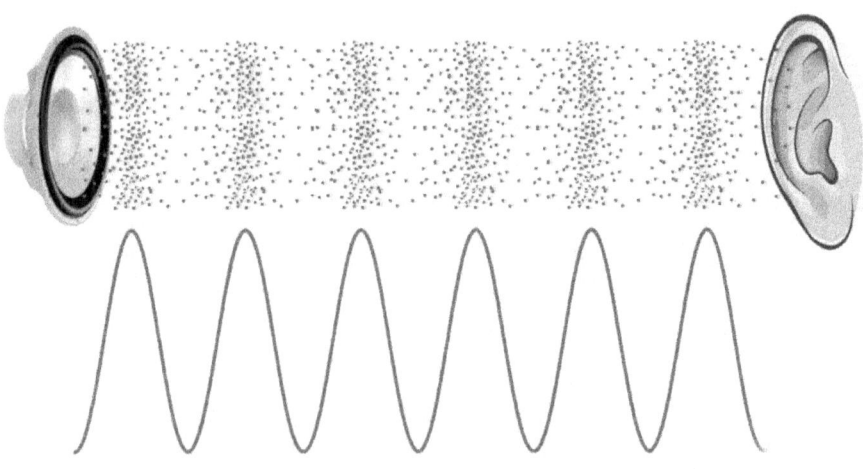

Sound travels in waves. Anything that
travels faster than sound is called supersonic.

Break the Barrier

In 1947, pilot Chuck Yeager flies
NASA's Bell X-1 plane faster than the
speed of sound. He is the first person
to break the sound barrier! Because
of the Cold War, the United States
keeps it a secret for years.

1947

At 768 miles per hour something strange happens. The plane is going as fast as sound. When the plane goes faster than the speed of sound, there is a sonic boom. This makes the plane hard to control. The sonic boom uses lots of energy, so planes fly inefficiently too.
Flying faster than sound is more expensive than flying at speeds below the sound barrier.

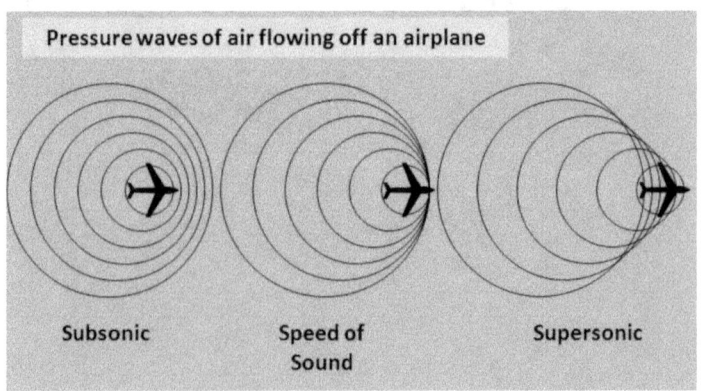

Pressure waves of air flowing off an airplane

Subsonic   Speed of   Supersonic
                  Sound

Planes — Past and Present

Sky Spies

The SR-71 Blackbird Spy Plane flies over three times the speed of sound, also called "Mach 3." It flies at an altitude higher than 60,000 feet.

Even though the plane is high in the air, it takes clear pictures of objects on Earth below.

For example, from 10 miles (16 kilometers) above, it can read the license plate of a car on the ground.

The SR-71 Blackbird flies at 3 times the speed of sound.

Planes — Past and Present

Sneaky Stealth

Stealth planes hide from radar. This
makes the planes harder for enemies to find.

F-117 Nighthawk

With science, stealth planes have special shapes
that deflect or absorb radio waves. This makes
these planes harder for radar to detect.

B-2 Spirit

Notice there is no tail. Digital computers help control this plane.

SIX) Compete
Commercial Competition

The jet engine has non-military
uses too. Business people create
commercial planes to fly paying passengers.

There is stiff competition between the
companies and countries that make planes.

Boeing

Boeing names their planes in
the 700 series like 727 and 737.

737

Wikipedia by EyOne

747

Planes — Past and Present

McDonnell Douglas

McDonnell Douglas names their passenger planes DC, for Douglas Commercial. There is the DC-6,7,8,9 and 10. In 1997, the McDonnell Douglas company merges with The Boeing Company.

Airbus

European Airbus names their planes A3XX like the A330 and A340. The governments of Germany, France, England and Spain work together to make Airbus planes.

Planes — Past and Present

Fuller and Farther

In the 1970s, Boeing makes the 747.
It carries four to five hundred people.
It can fly over 8,350 miles (13,450
kilometers) without refueling.

The 747 is a "fly-by-cable" plane. The pilot in the cockpit is directly connected by cables to the flight controls.

Flight Control Cables

The Boeing 747 used jam-proof cables to control flaps and actuators directly from the cockpit.

Planes — Past and Present

Supersonic Concorde

The Supersonic Concorde flies at twice the speed of sound. It flies from London to New York City in about three hours. Only twenty planes are made.

Supersonic means faster than the speed of sound. The sound shock wave causes sonic booms, which take a lot of energy to overcome. Because the Concorde planes are expensive, they only fly for 27 years. The Concorde is an amazing engineering feat.

The name comes from the French word "concorde" which means agreement. It symbolizes how the French and British work together to make this plane.

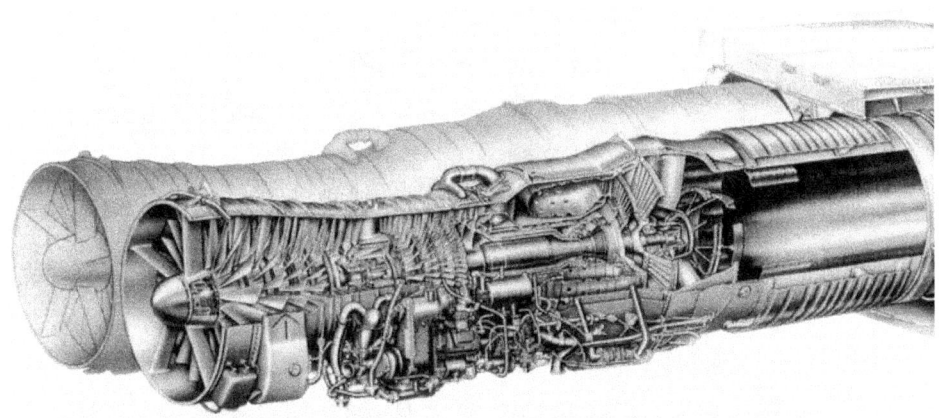

Concorde airplanes fly from 1976 to 2003.

Wikipedia by Russavia

Notice, when the Concorde is on the ground,
its nose points down, so the pilots can see better.

Konkordski

The Soviets make the TU-144 supersonic passenger airplane. It looks suspiciously like the Concorde. Some think the TU-144 design is stolen from Concorde. They nickname this plane "Konkordski."

In 1978, the TU-144 crashes. The passenger fleet is permanently grounded after only 55 flights.

They build seventeen TU-144 supersonic airplanes.

Fly-By-Cable Recap

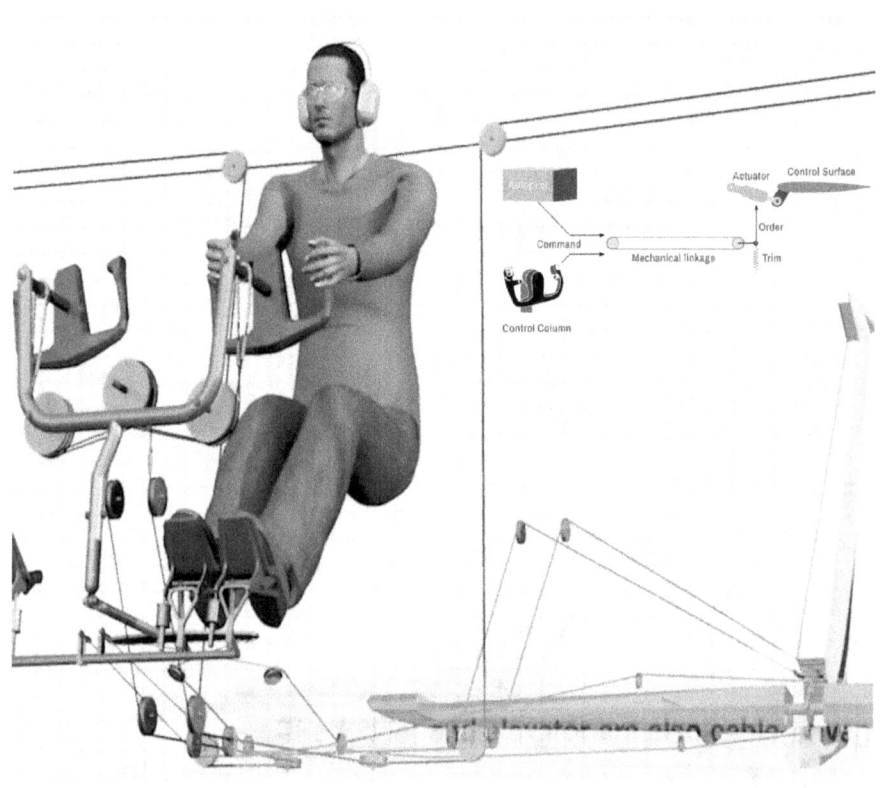

To recap, up to the 1980s, most planes are "fly-by-cable." That is, the pilot moves the wheel and rudder pedals. Cables connect directly to the flight control surfaces such as flaps, slats and rudders. All 737s are fly-by-cable.

Fly By Wire

Today, most planes are "fly-by-wire." Bits
of electricity (e-Bits) and actuators replace
mechanical flight controls. The pilot moves
sensors that send e-Bits to digital computers
through wires. More wires connect the
computers to actuators that physically
move the flight controls flaps.

Super Jumbo

Europeans make the Airbus A-380. It is the world's largest passenger plane. It has double decks and holds more than 600 passengers. It is nicknamed the "Super Jumbo."

Boeing 787

Instead of competing with Airbus for the world's largest plane, Boeing chooses to make a more efficient plane, the 787. Most of the 787 body is made of light carbon composite materials instead of metal.

Most of the 787 body is made of light composite materials.

787 window dim

Planes — Past and Present

SEVEN) Concepts
Future

What is the future of human flight? Will planes be electric
or fusion powered? Will planes go hypersonic speeds? About
200 years ago, the first passenger trains enter service.
At that time some people think that if humans go
faster than 20 miles per hour then heads might explode.
Thankfully, this is proven false. This may be the source
of "loco" (Spanish for crazy) in the word "locomotives"
which means "vehicles for pulling trains."
Hypersonic means more than five times the speed of sound.
Hypersonic will have mega tech challenges to achieve.
What will humanity do with all the hyper-speed? Will
people use it to enable weapons and wars or to enact
global peace and universal prosperity? Peace is worth a try!

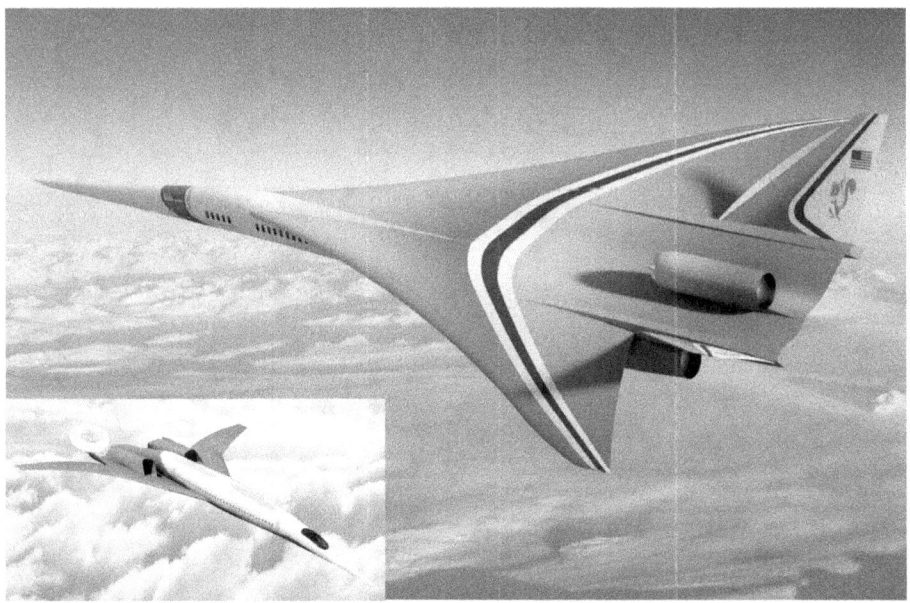

Ansari X Prize

Efficiency is important but the aerospace dream is still, faster, farther and better. In 1996, the "X Prize" offers $10 million to the first private vehicle to fly to the edge of space. That is, to fly 60 miles (100 kilometers) high. In October 2004, SpaceShipOne wins the X Prize. It costs Paul Allen, the co-founder of Microsoft, $100 million to achieve it. New technologies continue to be invented to take humans to the end of the sky.

SpaceShipOne is the first private plane to go
to the edge of space at 60 miles (100 kilometers) high.

Sky's No Limit

Watch an airplane take off.

It all seems so elegantly easy.
There are spirits, prizes and
sonic booms to get us to this point.
Airplanes now circle the globe daily.

The story of flight is a tale of
human technology and triumph.

As great as flying accomplishments are,
it makes us wonder — is the sky but a
springboard to outer space?

Planes — Past and Present

In the 100 plus years that passenger planes are flying, people fly over 100 billion miles (160 billion kilometers).

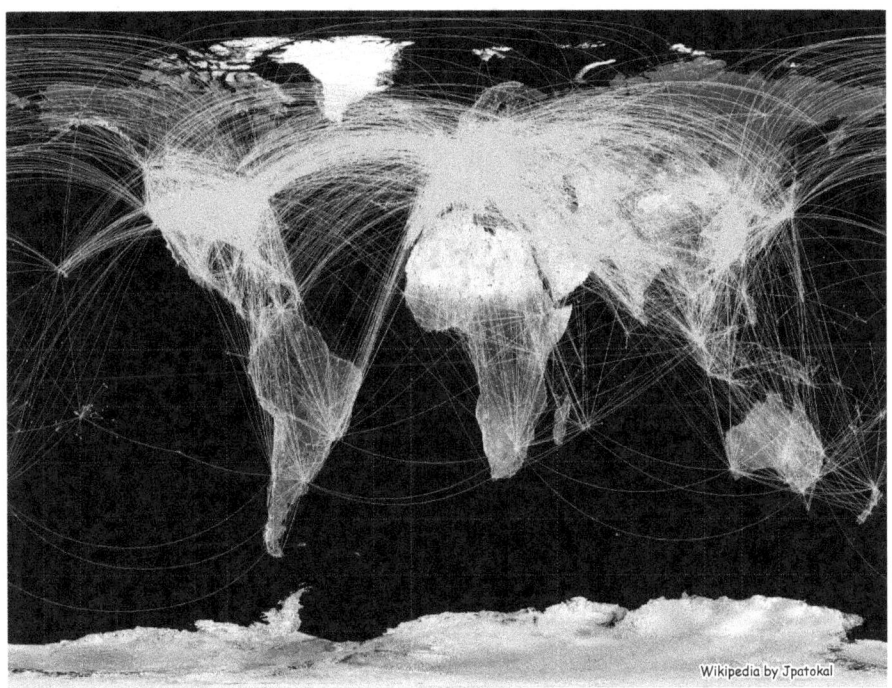

Wikipedia by Jpatokal

What starts long ago by seeing birds flapping, is now humans flying around the world. With the science of jet engines and curved wings, we earthlings soar to the skies. Our feet are no longer gravity-glued to the ground. We are sky-lings now!

Credits

Unless otherwise noted, pictures are in the public domain or Fair Use.

We would like to thank:

. www.wikipedia.org

. Smithsonian's National Air and Space Museum

. Boeing Museum of Flight

. London Science Museum

ONE PAGER

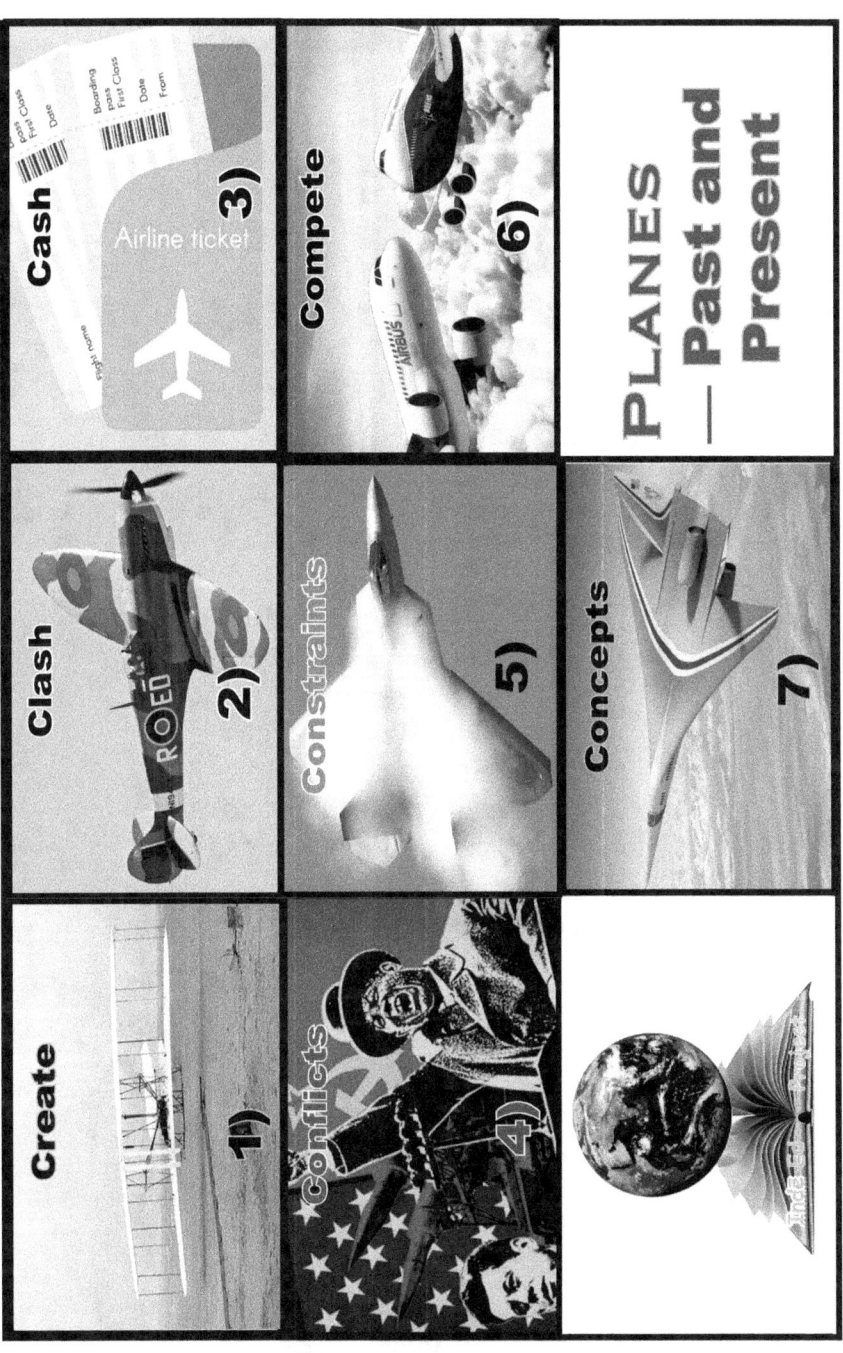

Create 1)

Clash 2)

Cash 3)
Airline ticket

Conflicts 4)

Constraints 5)

Compete 6)

Concepts 7)

PLANES
— Past and
Present

Watch VIDEO

Main Points
1) Create first flights
2) Flying machine clashes
3) Pay cash to fly
4) Cold War conflicts
5) Speed constraints
6) Countries & companies compete
7) Future of flight concepts

For a copy of this video contact:

To Recap

From bird flaps to flying planes,
we learn about four forces.
Air is important to why planes fly too.
When we understand science, we soar!

Easy Ideas	Airplanes	Cars
From Concepts to Critical Thinking	From Four Forces to Flights	From Actions to Autos
1	2	3
Computers	Smartphones	Food
From Digital to Data	From Calls to Global Connects	From Eats to Energies
4	5	6
Nature	Space	Light
From Atoms to All Life	From Elements to Us	From Suns to Sapiens
7	8	9
AI	STEM-Zen	Bonus Everyday Objects
From Machine Muscles to Minds	Program From Empty to Science EnLights	From Ideas to Daily Items
10		Planes — Past and Present

What Is It?

The STEM-Zen Program

is an integrated SCIENCE Program with thousands of pages and over 50 videos. Teachers help students go from science empty to knowledge enLighted!

2) Airplanes
— Past and Present

How do we go from watching birds flapping their own wings to flying ourselves? This book highlights this off-the-Earth adventure.
With jet engines and curved wings, we earthlings become airborne. Our feet are no longer gravity-glued to the ground. We are sky-lings now!

2) Airplanes
— Wedge Tools

There are links
between wedge-shaped
tools and jet wings.

Douglas J. Alford
Sally Kimangu

STEM-Zen Program

Wedge Tools

Table of Contents

Without tools,
hands work but weakly.
With science, we build machines
and harness powers to enable our creativity.

2 - WEDGE TOOLS One Pager

Let's see how science connects
from tool <u>wedges</u> to leading <u>edges</u>.

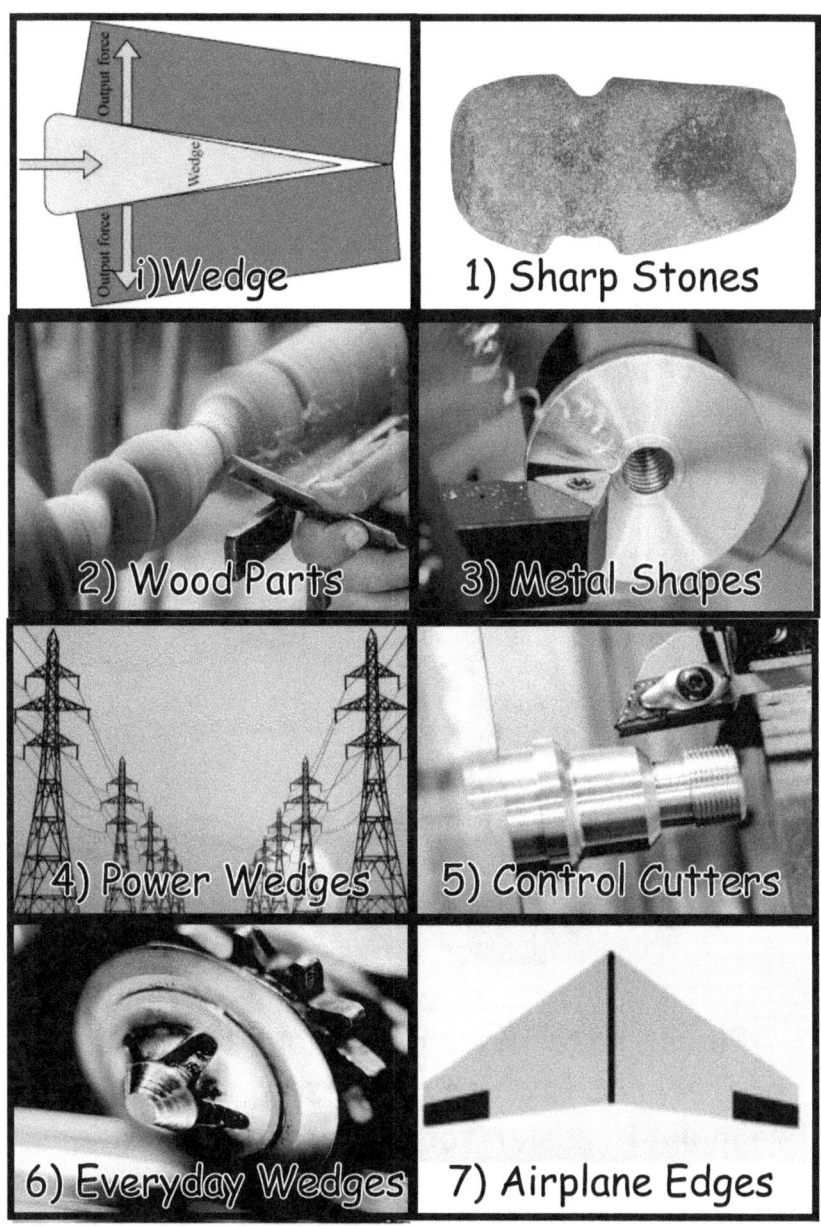

i)Wedge

1) Sharp Stones

2) Wood Parts

3) Metal Shapes

4) Power Wedges

5) Control Cutters

6) Everyday Wedges

7) Airplane Edges

In this chapter, watch for
the following scientific principles:

- Wedge
- Inclined Plane
- Cutting Edge
- Screw
- Gears
- Electricity
- Computer Numerical Control
- Machine Learning
- Propeller
- Leading Edge

Please note that throughout this book,
preference is given to present tense verbs
to make it easier for international readers.

The WEDGE is a "V" shape that splits things apart. Tips of tools are wedges that change our world.

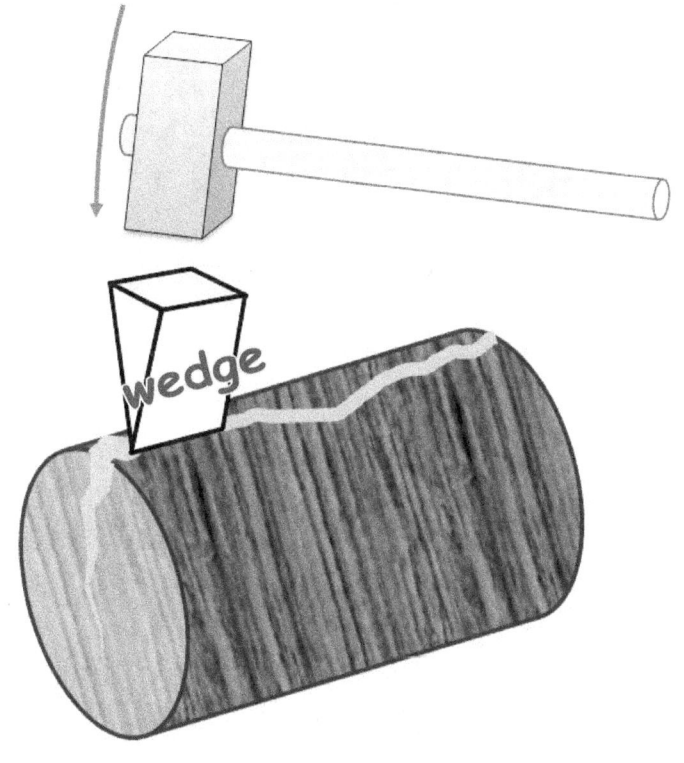

wedge

This book looks at snapshots of different wedges — like puzzle pieces — that fit together to make our modern world.

Our ancestors survive because of these cutting wedge-shaped tools.

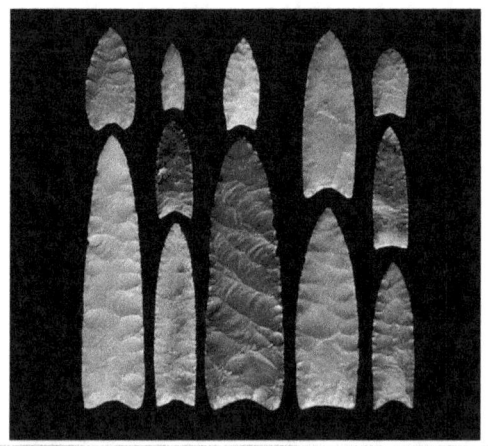

Hunting gives new meaning to the words "fast food."

Indeed, each step in human progress — from caves to computers — is enabled by inventing better tools.

computers and data centers

With science, people progress from stone to bronze and then iron tools.

Wedges

When we see everyday objects, may we wonder at the tools used to make them.

Wedges

Simple wedges connect through time to our complex technologies today.

Let's start our true story with V-shaped rocks.

ONE, Stone Wedges

Long ago, people put V-shaped stones on wooden handles to make axes. Axes chop wood to make our homes and get fuel to cook our food.

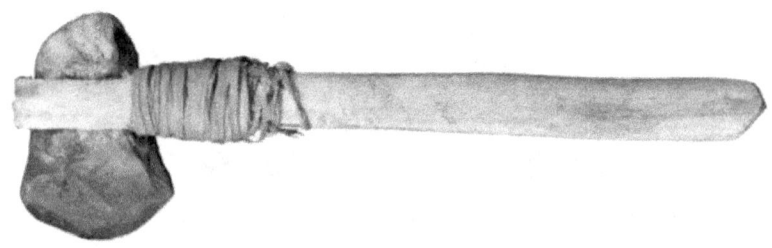

Next, people chip rocks
to make large spear points.
Cleverly, people work
together to hunt big animals.

obsidian

Also, people chip rocks to make short arrowheads. Bows launch the missile like arrows to hunt small animals. People use long, sharp stone knives to prepare the meat for eating.

Our ancestors use these
axes to make boats.
Wedge tools provide the ways
for people to move around.

 adze

Much later, people make wedge-shaped tools to plow soil, plant seeds and kill weeds.

Jethro's seed drill 1701

People, throughout time, turn wedges into weapons too.

Over time, people
discover different metals
to make better tools.
First, bronze then later steel.

Egyptians use chisel-tipped
bronze tools to make the pyramids.

Wedge-shaped ramps are used
to move the heavy stones.

Later, people make
strong steel tools.

Next, wedge tools shape wood.

TWO, Wooden Parts

Metal tools with sharp
wedge edges make useful
parts out of wood.

cutting edge

In the past, people use springy poles for power to turn round parts like wooden screws.

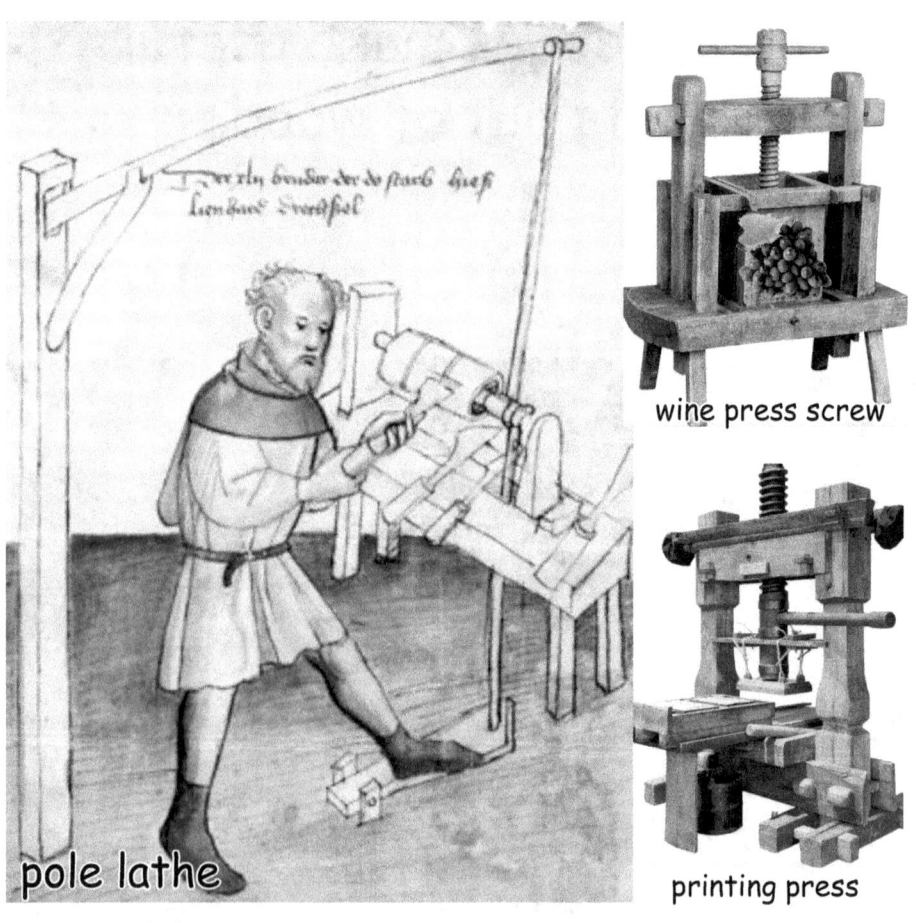

wine press screw

pole lathe

printing press

Wooden screws press grapes for wine. Wine press wooden screws are used on the first printing presses too.

Also, wedge-shaped saws
turn round trees
into rectangular lumber.

People make their homes and
household goods out of wood.

People use straight wedge-
edged tools to make
flat pieces of wood
for tables and benches too.

wood planer

wood table

wood bench

In the past, wedge-tipped tools make wooden homes that put roofs over our relatives heads.

Wedges

Before cars, horses pull humans in wooden carts and carriages. They are made with wedge-shaped tools.

In the 1500s, wooden ships enable the human Age of Exploration.

Four hundred years later, the first airplanes fly.

They are mostly made of wood and cloth.

Wedge-shaped tools
help make wooden parts
for early airplanes.

Next, more triangle-tipped tools.

Wedges

THREE, Metal Shapes

Over time, people invent better
wedge-shaped tools to
make more precise metal parts.

Wedges

People cut spiral threads
on long metal cylinders
to make accurate screws.
Screws are winding wedges.

With precise lead screws, people make better lathes. Lathes have strong, wedge-shaped tools that turn **ROUND** metal parts.

People make metal tools that cut precise **FLAT** shapes called planes.

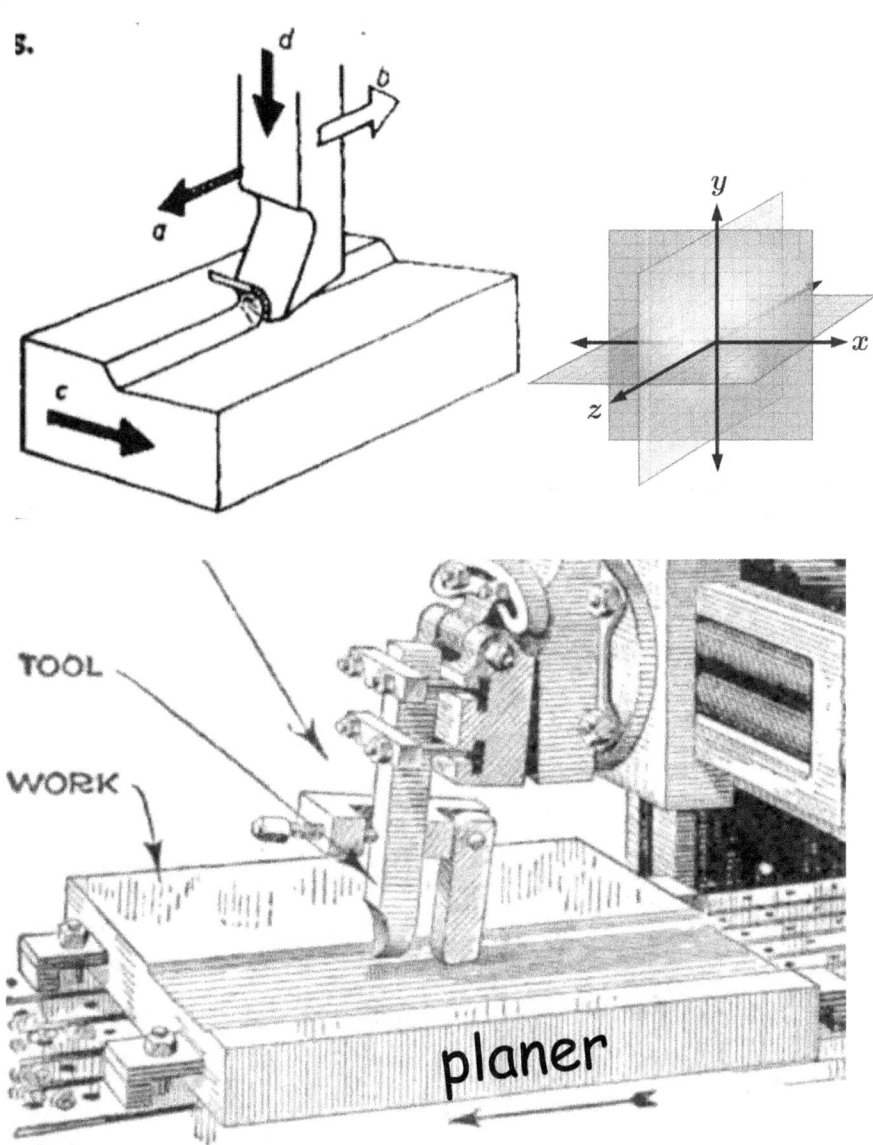

5.

TOOL

WORK

planer

These tools cut teeth into metal gears.

hobb

gear

Clocks, cars and airplanes use gears!

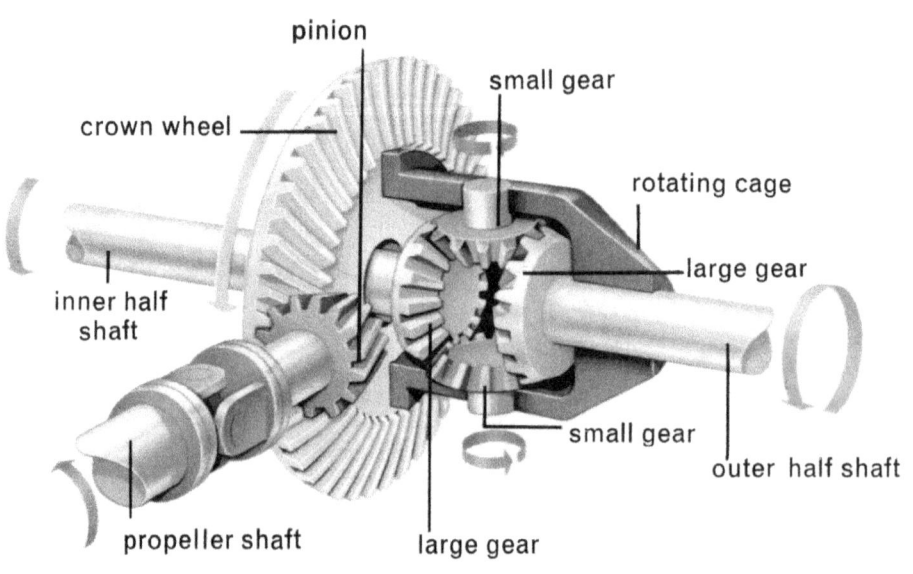

pinion

small gear

crown wheel

rotating cage

large gear

inner half
shaft

small gear

outer half shaft

propeller shaft

large gear

Wedges

Spiral tools called mills
make metal parts with
many different shapes.

Special whetstones and grinding wheels — with very tiny, hard, wedge-shaped grains — sharpen tools.

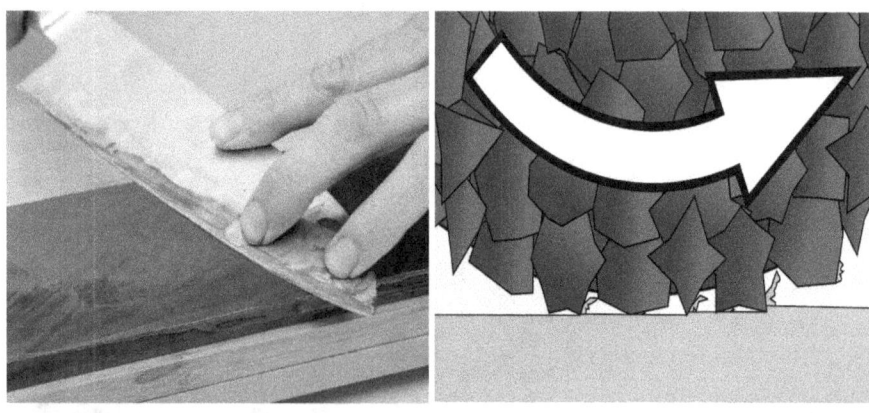

Next, wedges are energized!

FOUR, Power Wedges

Over time, people invent powered machines, with wedge-shaped tools that make useful parts.

electricity

Wedges

In the past, this machine
tool cuts cannon bores.
It also, makes accurate
cylinders for steam engines that
power an Industrial Revolution.

4-2

Today, metal lathes make
nuts and screws that hold
parts together. Also, lathes
make precise car parts like these.

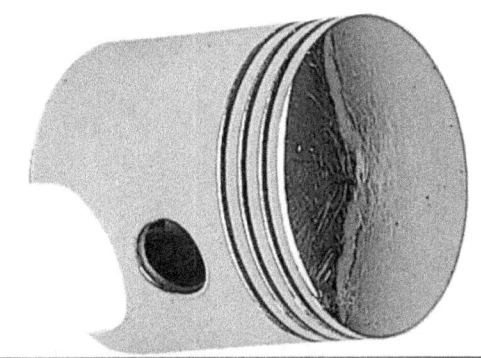

Car engine parts must be precisely flat, so they will fit together.

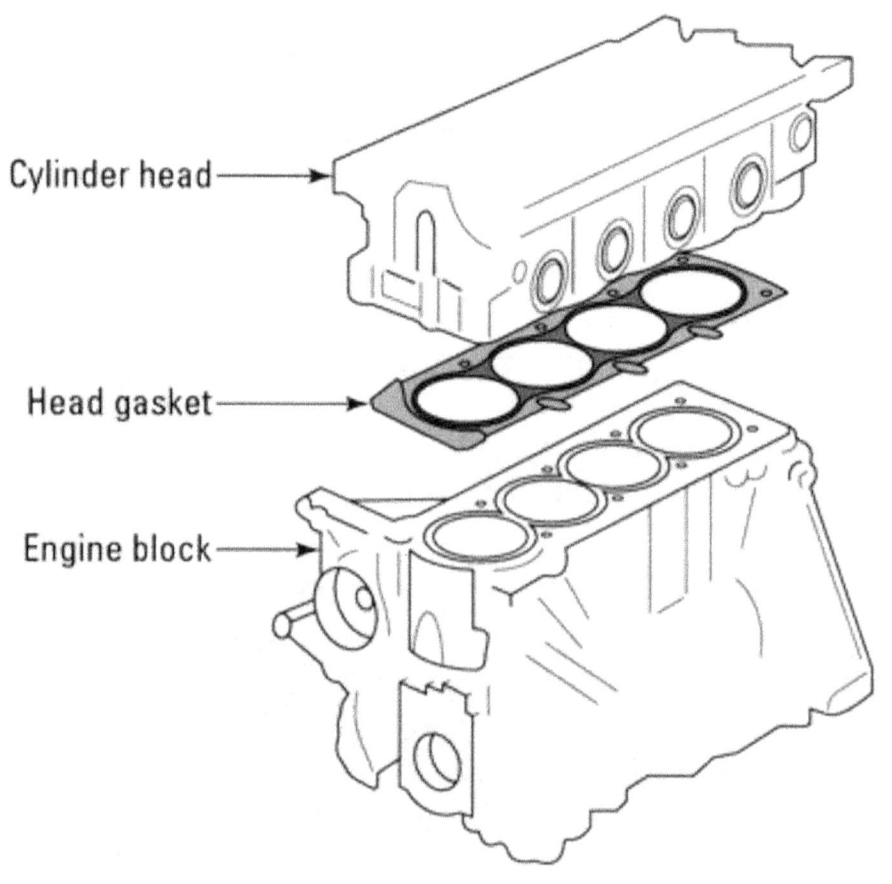

Cylinder head

Head gasket

Engine block

car engine block and head

Our world moves with gears made by wedge-shaped tools.

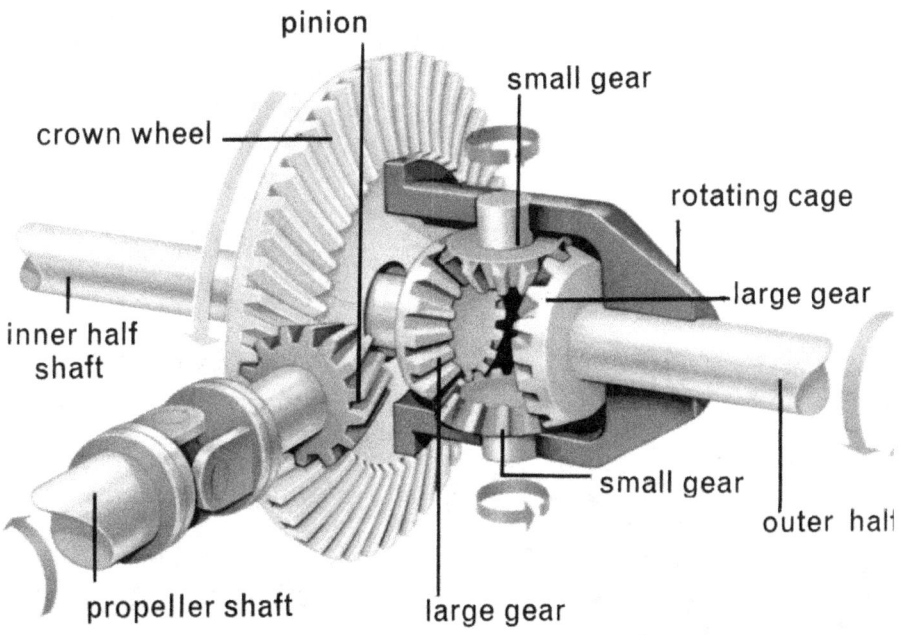

pinion

small gear

crown wheel

rotating cage

large gear

inner half shaft

large gear

small gear

outer half

propeller shaft

large gear

These gears transmit turning
from the engine to make tires move.

Electricity powers wedge-tipped mill tools that shape airplane parts like these wing spars.

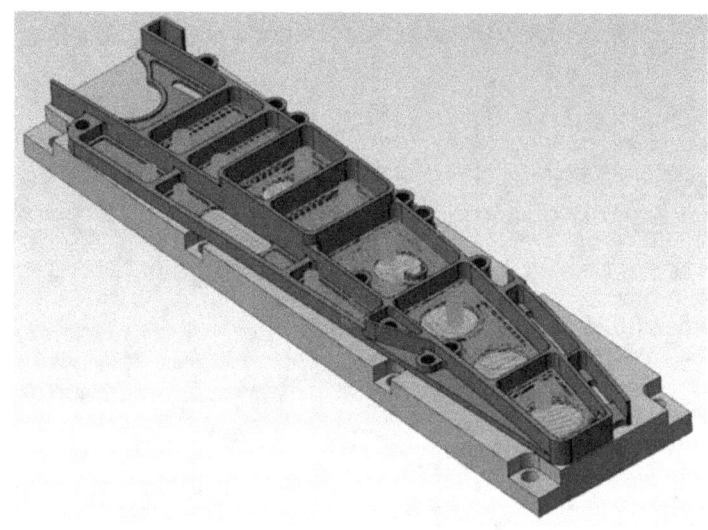

Tools have to be
carefully controlled to make
quantities of quality parts.

Wedges

FIVE, Controlled Cutters

At first, skilled people
control the machine
tools to make parts.

Wedges

Next, people invent motors
to move tools that are controlled
by numbers and math. People
program these early machine
tools with punch tapes.

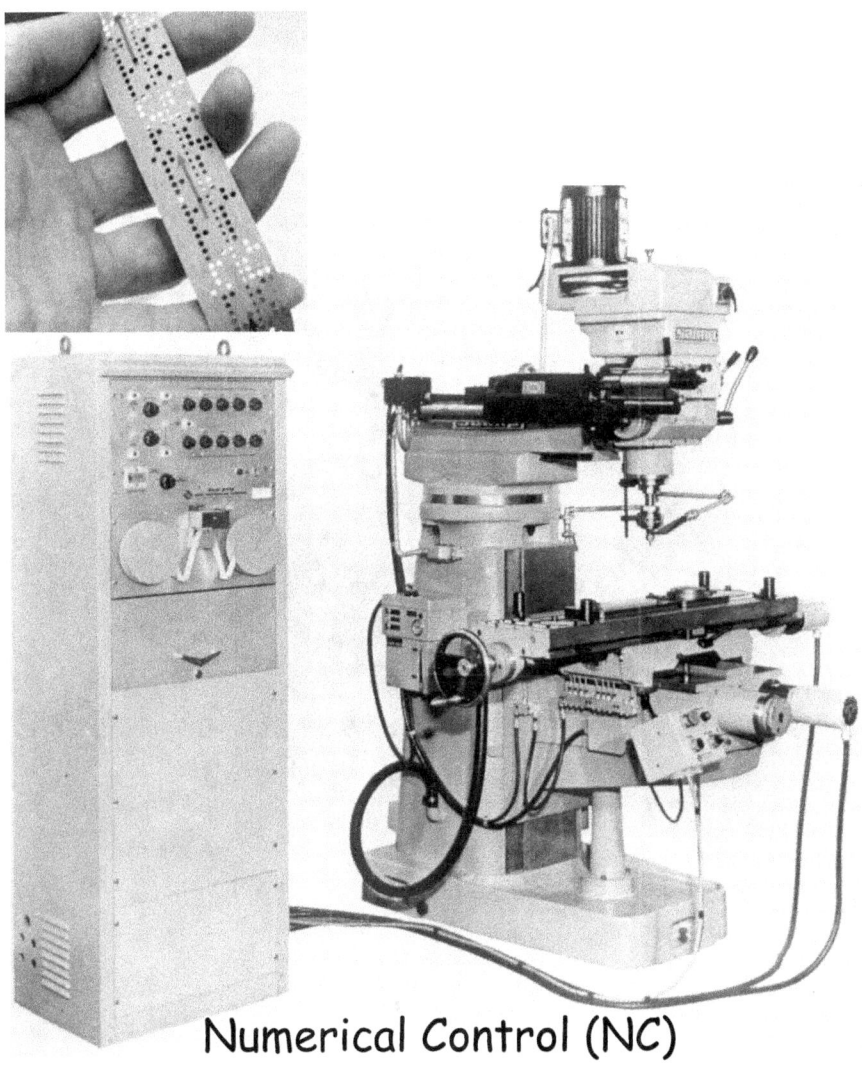

Numerical Control (NC)

Later computers are improved. People write software to control the machine tools too.

Today, computer controlled machines have many wedge-shaped cutting tools.

Computer Numerical Control (CNC)

CNC machine tools make many of our everyday things!

Simply said, withOUT wedge tools, there will be no cars or most of the everyday objects we all use.

People are currently working on ways
for computer machine tools
to self-learn how to make parts.

This is called "machine learning."

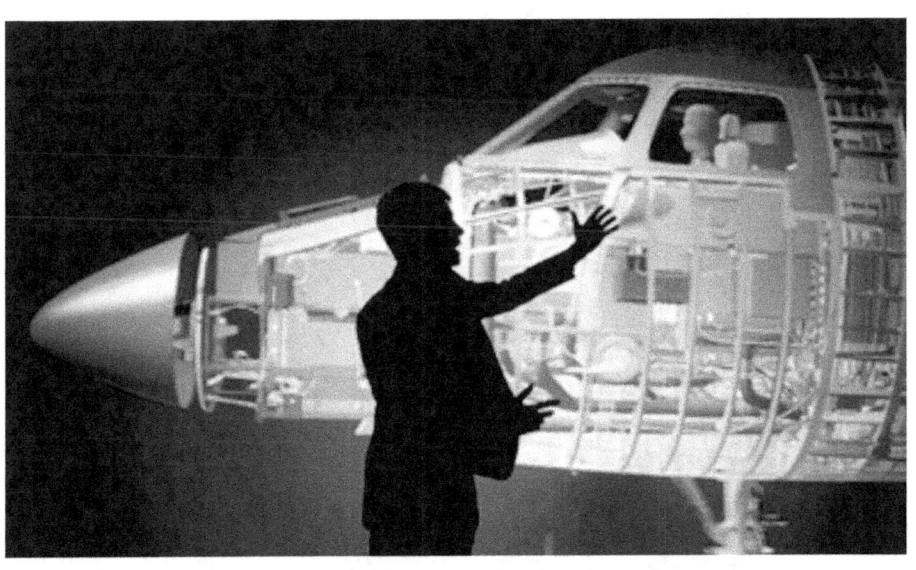

Next, more wedges.

SIX, Everyday Wedges

Wedge-shaped objects are all around us.

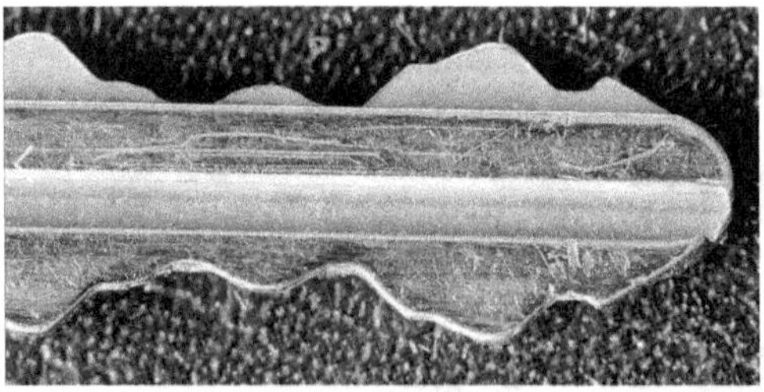

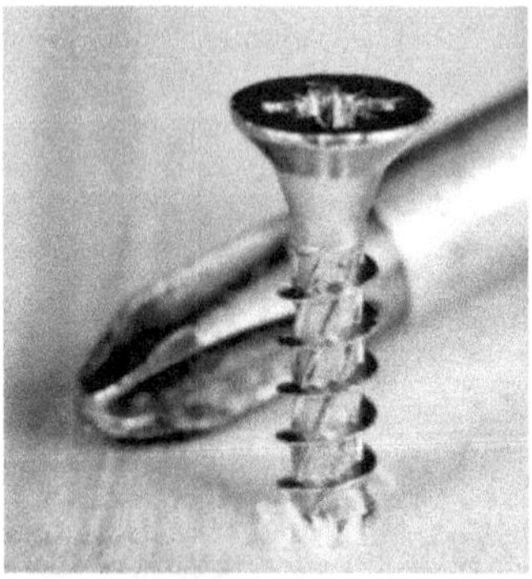

Our teeth are wedges.
Gear teeth are too.

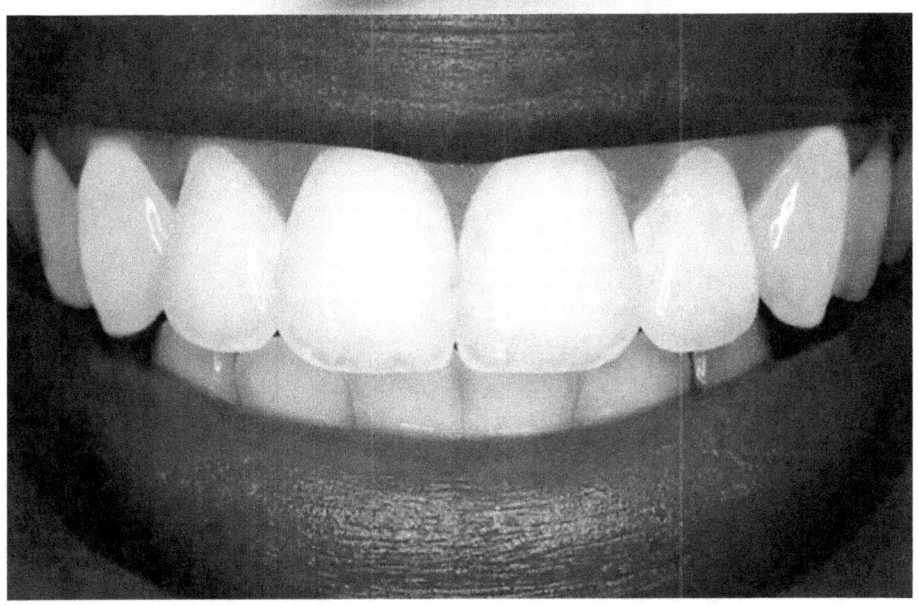

These garden hand tools are wedges.

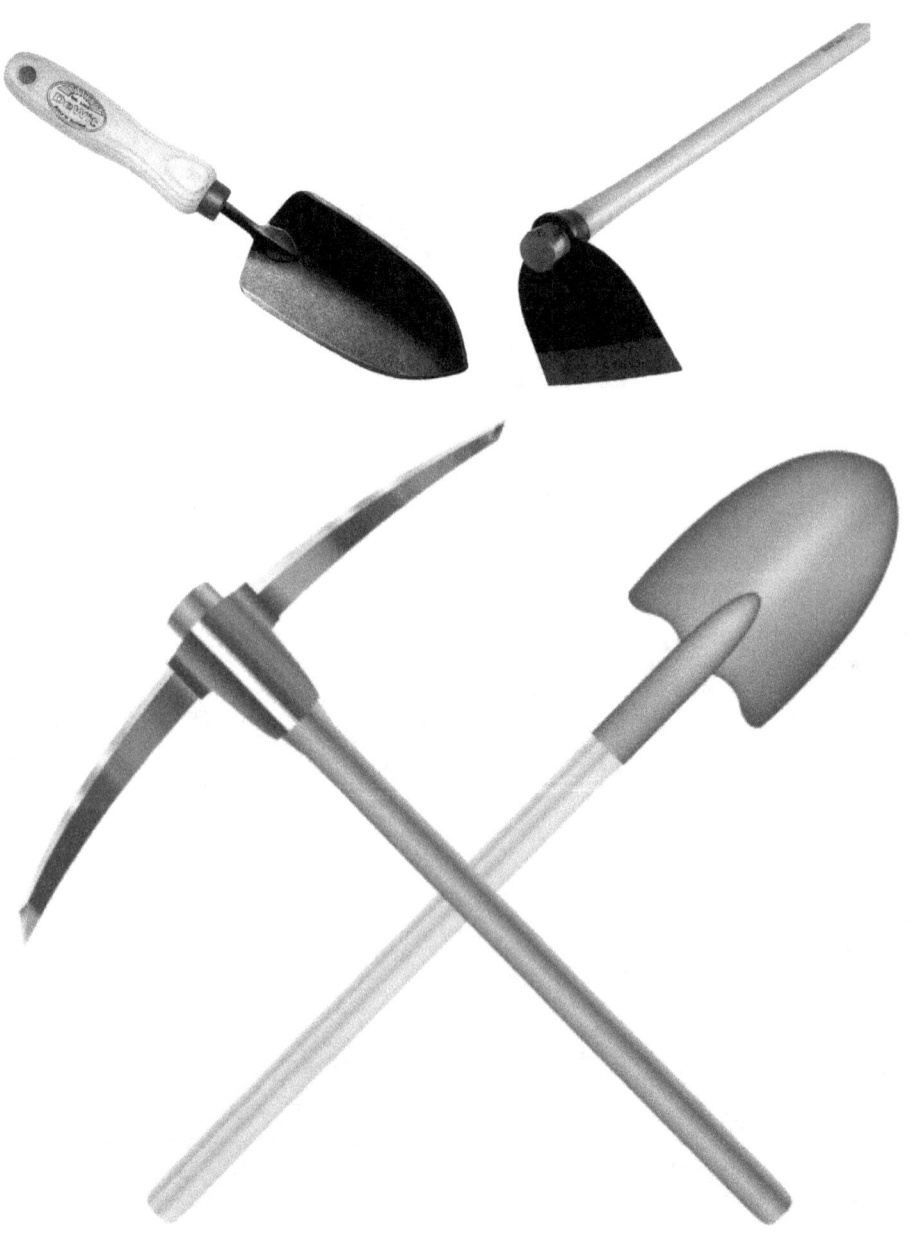

Can openers, scissors and zippers are wedges too.

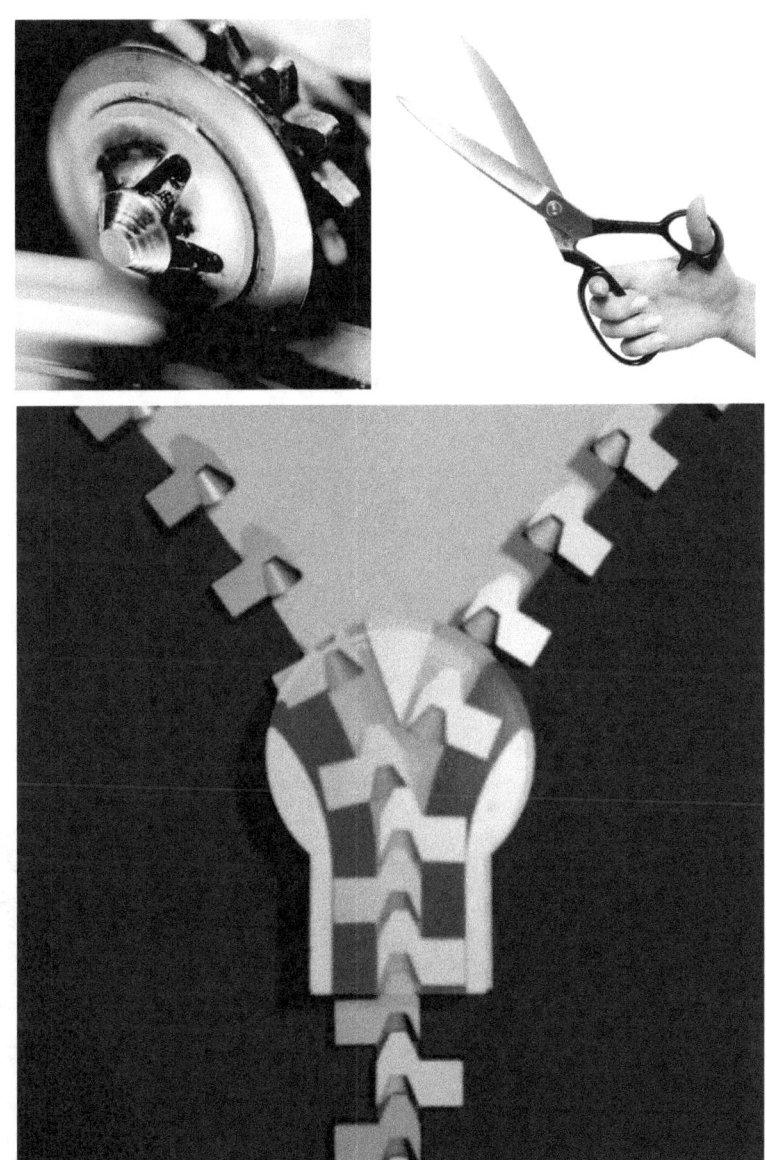

Like water pumps, wedge-shaped
ship propellers push water backwards.
This pushes the ships forward.

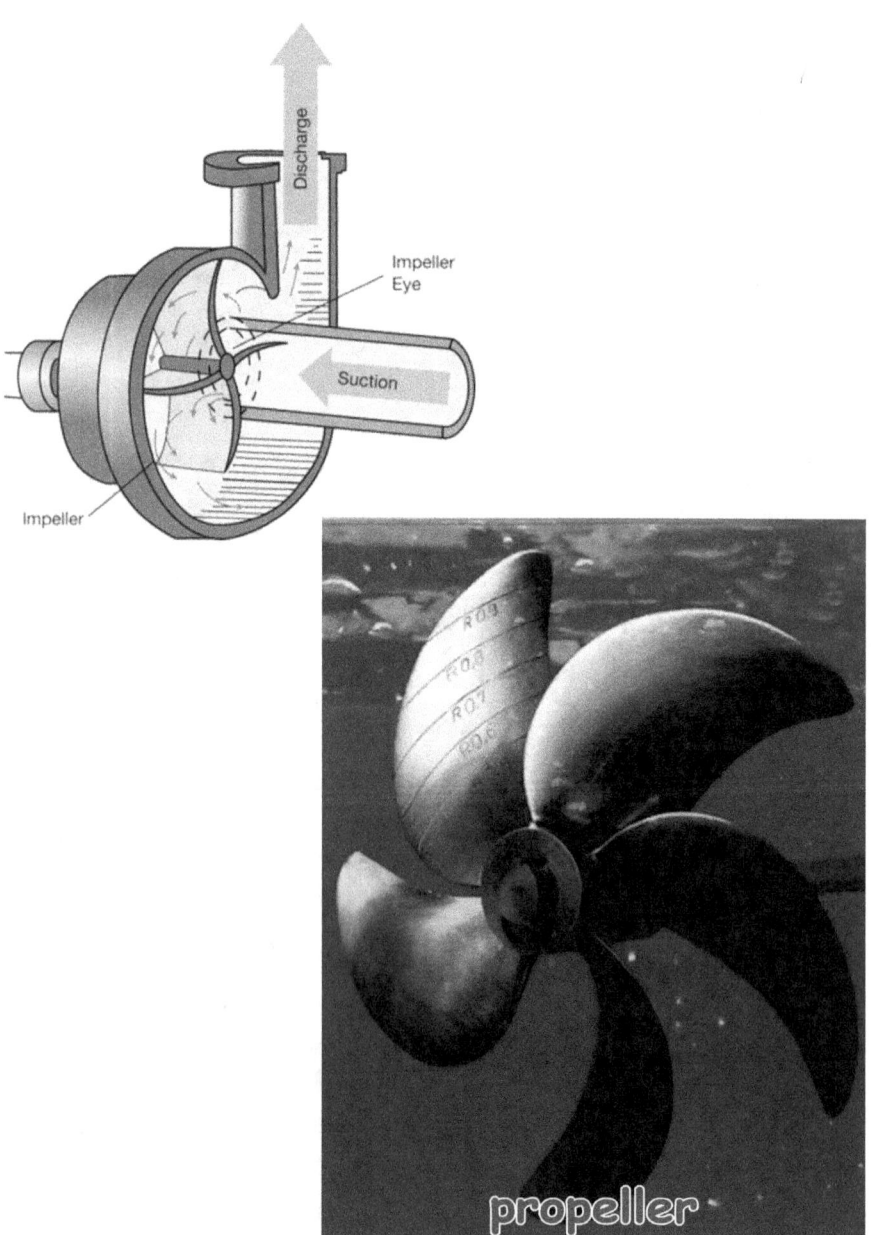

Our complex, high-tech
world has many objects made
by simple wedge-shaped tools.

Wedges

Which leads us
to wedges on wings.

swept wings

Wedges

SEVEN, Airplane Edges

The first planes have propellers that are twisting, turning wedges. They cut through the air and push it back to thrust the planes forward.

Propeller or "prop" planes
have straight wings.
Wings have wedge-shaped
cross-sections. This causes
air pressure to push the
planes up, called "lift."

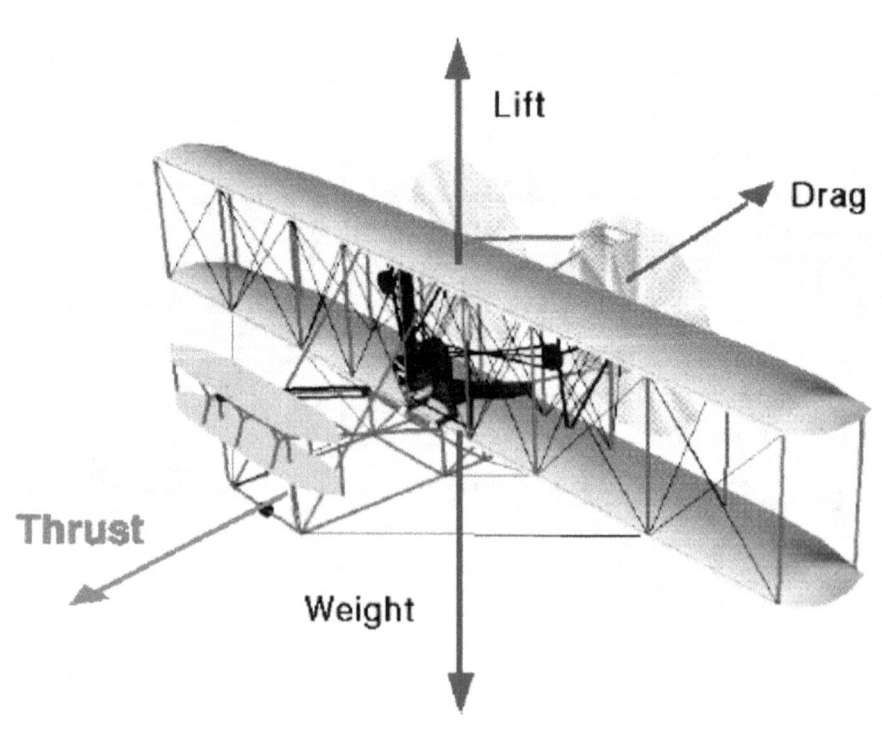

Today, jet engines have fan blades. Can you guess their shape? Yup! Wedges. They push together or compress air into the engine. More air or oxygen causes the fuel to burn better.

Why do jets have wedge-shaped wings?

Powerful jet engines cause the air to move strongly over the wings. This can cause the wings to vibrate and break apart. This is called "flutter."

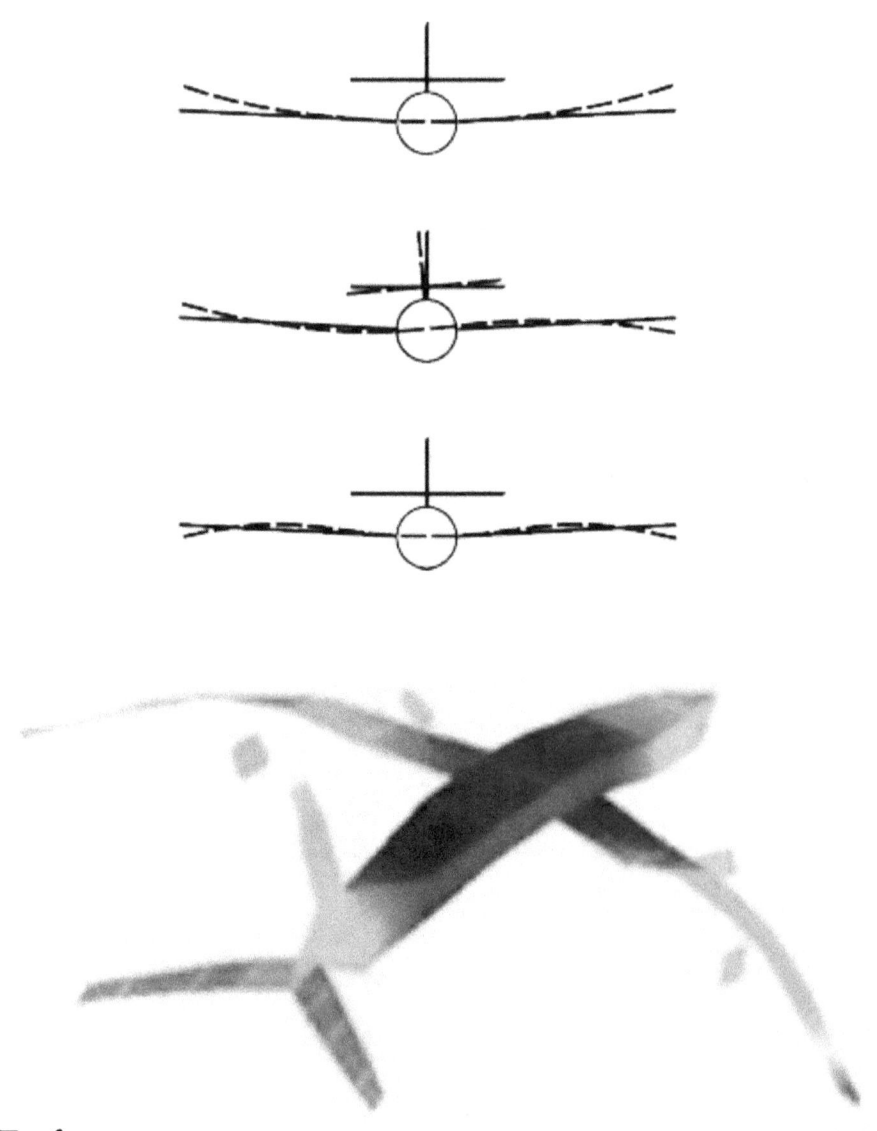

Thankfully, air flows smoothly over the swept-back, wedge-shaped wings with no flutter. Passenger jets have engines that are mounted on wedge-shaped wings.

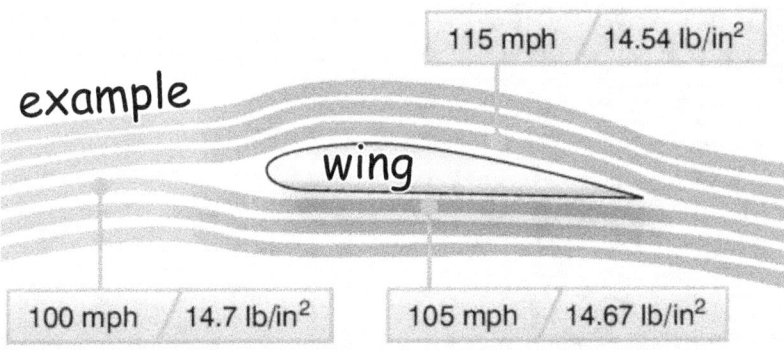

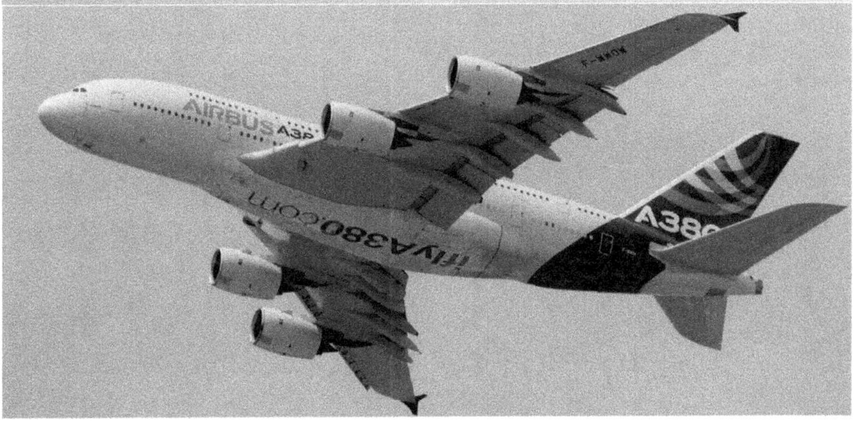

115 mph
14.54 lb/in²
example
wing
100 mph
14.7 lb/in²
105 mph
14.67 lb/in²

BOEING
787

AIRBUS
flyA380.com

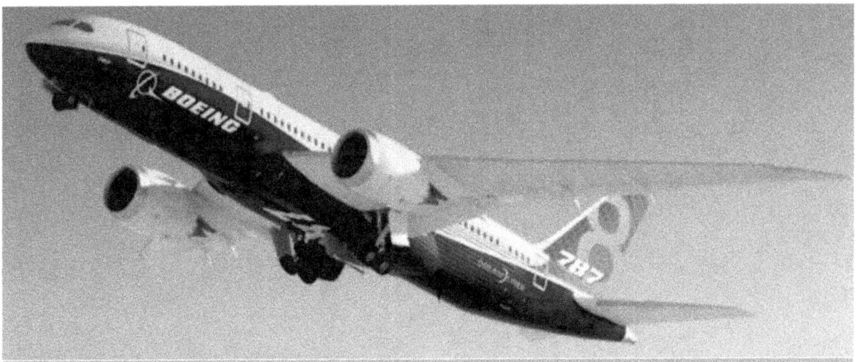

Flight control surfaces that steer the airplane are wedge-shaped.

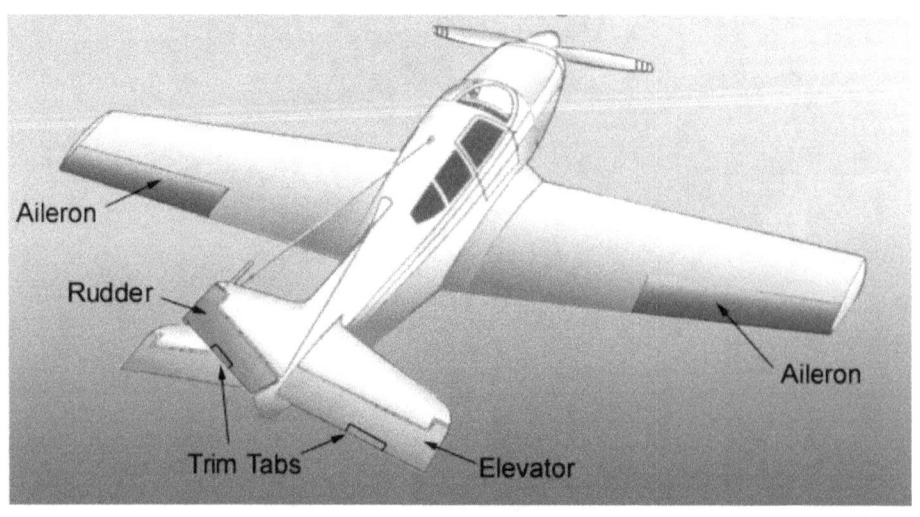

Aileron

Rudder

Aileron

Trim Tabs

Elevator

On airplane wings, the
important front face
is called the "leading edge."
The idiom, "cutting edge or
leading edge" means the "latest
in technology" too. All this
starts with wedge tips.

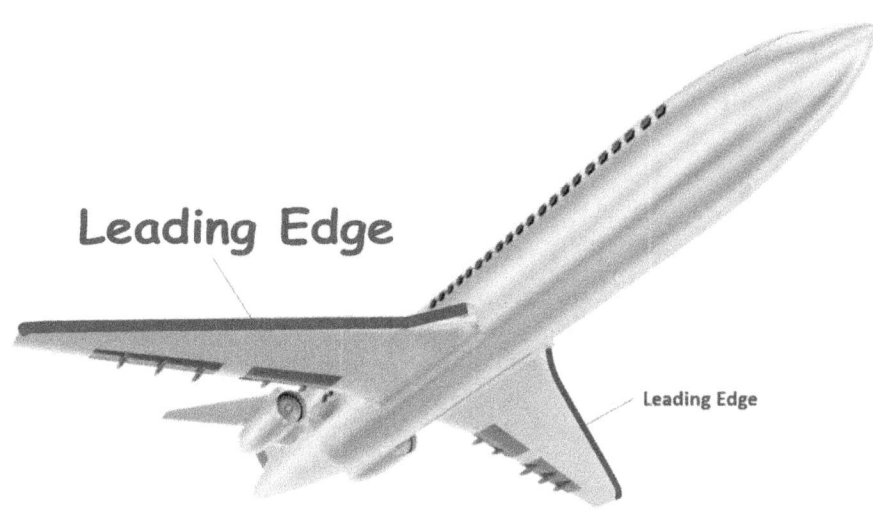

Leading Edge

Leading Edge

To close, ancient people
used sharp rock wedges
for axes and arrows.

Next, sharp metal
tools cut wood for our
homes, carts and ships.

Wedges

Later, stronger tools shape metal parts too.

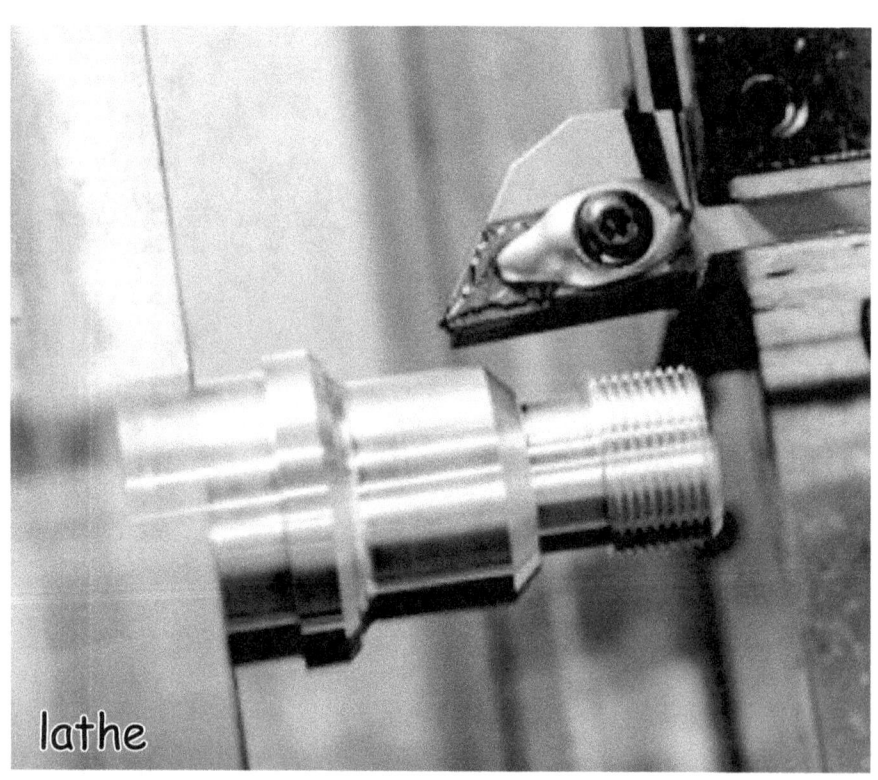

lathe

Wedges

These are examples of wedges that are all around us.

gears

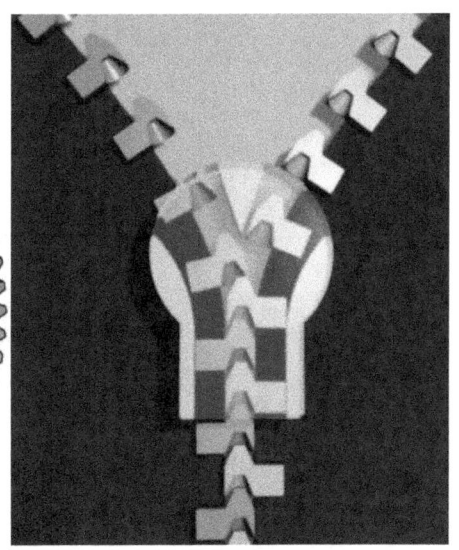

zipper

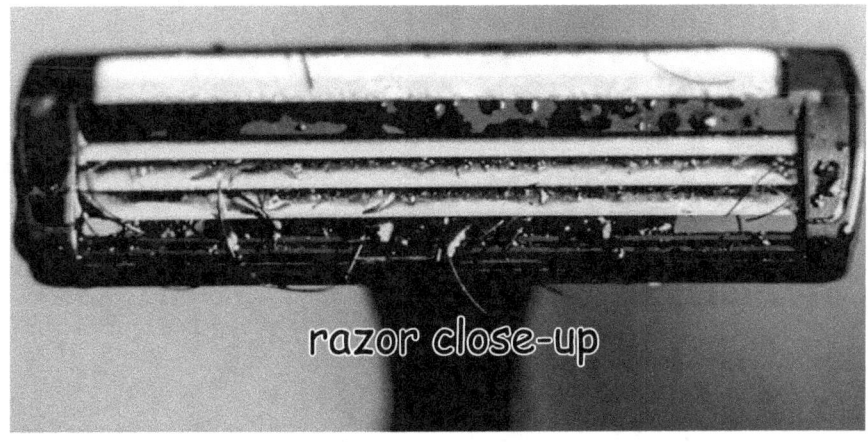

razor close-up

Today, computers control machines with wedge tools that make many of our everyday objects. This includes our high-tech items too.

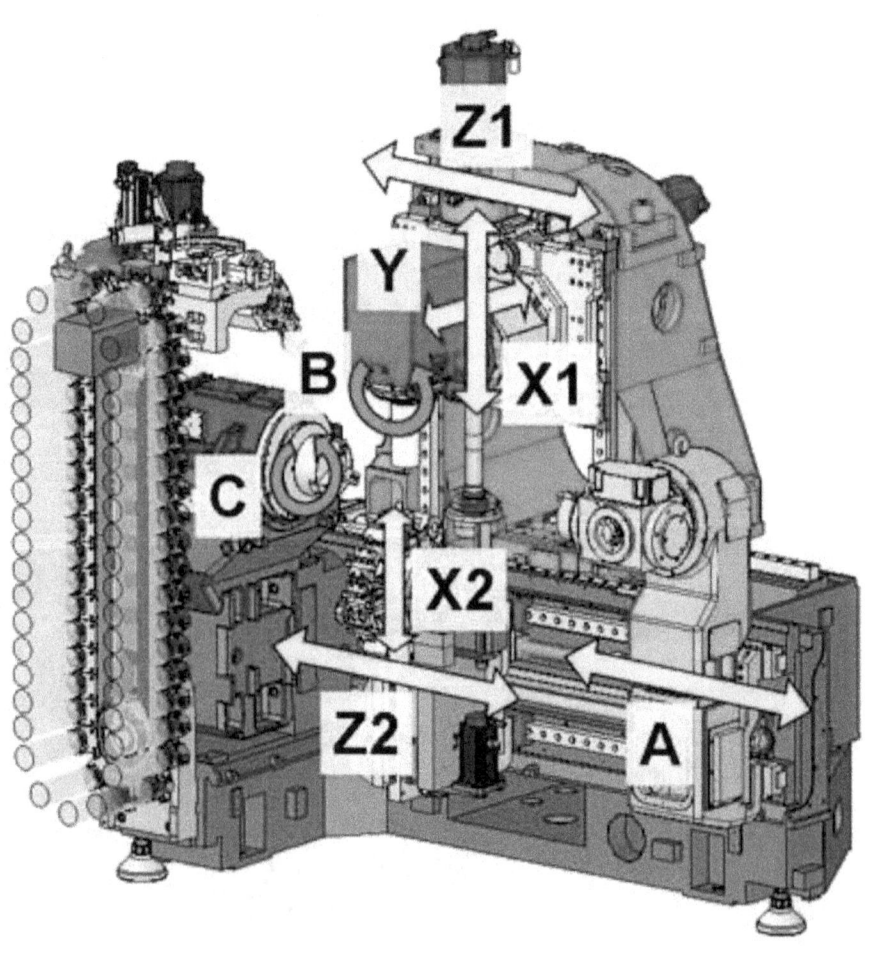

Also, wedge-shaped tools make many airplane parts. Jet planes even have wedge-shaped wings.

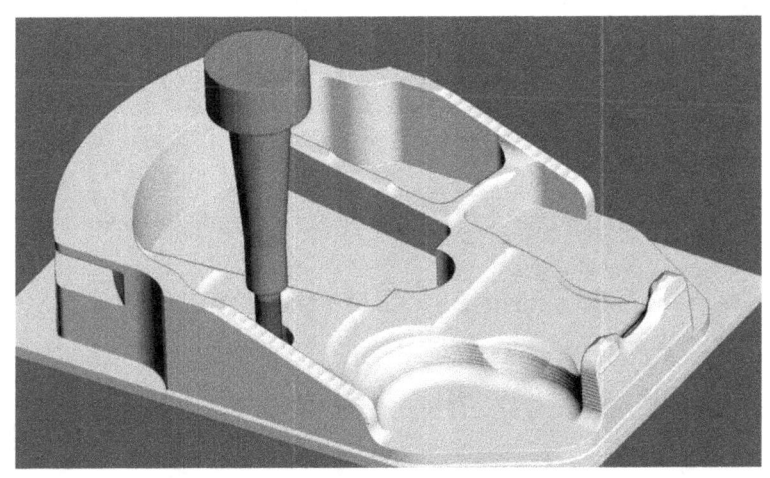

When we know about wedges, we better understand the science inside the cutting edge tech of our world today!

At the tips of cutting wedges, science changes human lives and enables leading-edge technology.

WEDGE TOOLS One Pager

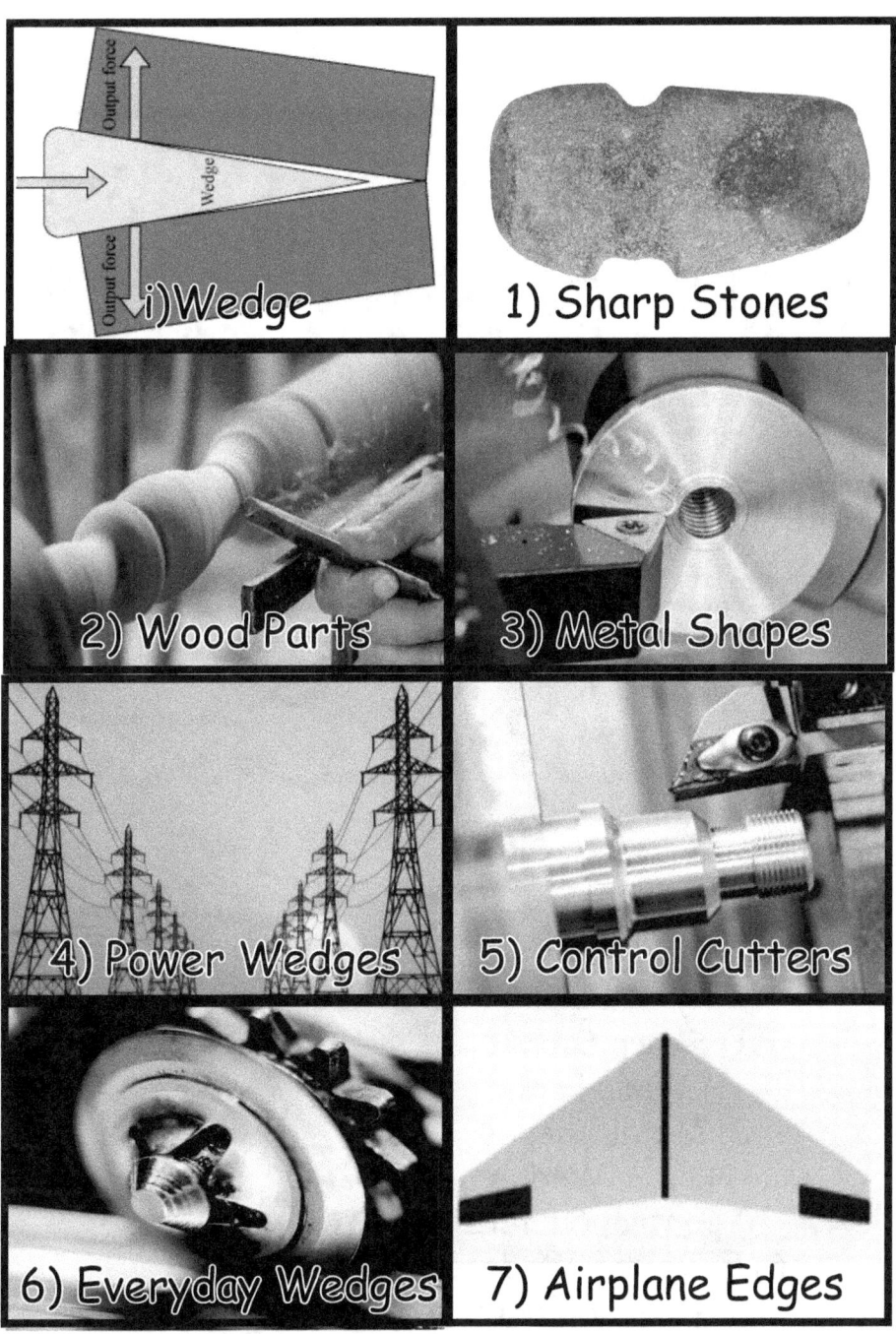

i) Wedge

1) Sharp Stones

2) Wood Parts

3) Metal Shapes

4) Power Wedges

5) Control Cutters

6) Everyday Wedges

7) Airplane Edges

Wedges

Watch VIDEO
Cutting Wedges

For a copy of this video contact:

<u>Main Points</u>
i) Wedges
1) Sharp Stones
2) Wood Parts
3) Metal Shapes
4) Power Wedges
5) Control Cutters
6) Everyday Wedges
7) Airplane Edges

2 - Cutting Wedges
-Axes to Airplanes VIDEO!- Script

i-i-1) Welcome to Tech Links - Cutting Wedges

i-2) Our ancestors survive because of cutting wedges-shaped tools.

i-3) Indeed, Each step in human progress - from caves to computers
- is enabled by inventing better tools!

i-4) With science, people progress from stone to bronze and then iron tools.

i-5) When we see everyday objects, may we wonder at the tools used to make them.

i-6) Simple wedges connect through time to our complex technologies today.

i-7) Let's start Our true story with "V" shaped rocks.

1-1) ONE, Stone Wedges Long ago, People put V shaped stones on wood handles
to make axes. Axes chop wood to make our homes and get fuel to cook our food.

1-2) Next, people chip rocks to make large Spear points.
Cleverly, people Work together to hunt big animals.

1-3) Also, people chip rocks to make short arrowheads.
Bows launch the missile like arrows to hunt small animals.

1-4) Our ancestors use these axes to make boats to get around.

1-5) Later, people make wedge-shaped tools to: plow soil; plant seeds; and kill weeds.

1-6) People throughout time turn wedges into Weapons too.

1-7) Over time, people discover different metals for tools. Egyptians use bronze tools
to make pyramids. Later, stronger steel tools are made. Next, wedges shape wood.

2-1) TWO, Cut Wood Metal tools with sharp wedge edges make wooden parts.

2-2) People use springy poles for power to turn round parts like wooden wine screws.

2-3) Also, wedge-shaped saws turn trees into lumber.

2-4) People use straight wedge-edged tools to make flat pieces of wood like tables and
benches.

2-5) In the past, wedge-tipped tools make wooden homes that put roofs over our relatives
heads.

2-6) Before cars, horses pull humans in wooden carts and carriages.

2-7) In the 1500s, wooden ships enable the human age of exploration.

2-8) 500 years later, the first airplanes fly. They are mostly made of wood and cloth.

3-1) THREE, Metal Shapes Over time, people invent better wedge
shaped tools to make more precise metal parts.

3-2) People cut spiral wedges on long metal cylinders to make
accurate screws. Screws are winding wedges.

3-3) With precise lead screws, people make better metal lathes.
Strong, sharp, wedge cutting tools turn ROUND metal parts.

3-4) People make metal tools that cut precise flat shapes called planes.

3-5) These tools cut teeth into metal gears for clocks.

3-6) These spiral tools called Mills make many different shapes.

3-7) Special whetstones and grinding wheels - with very tiny, hard,
wedge-shaped grains - sharpen tools.

3-8) Next, we energize wedges. 4-1) FOUR, Power Wedges
Powered machines move wedge-shaped tools to make parts that people use.

4-2) In the past, this machine tool cuts cannon bores. It also, makes accurate
cylinders for Steam Engines that power an industrial revolution.

4-3) Today, metal lathes make nuts and screws that hold parts together.
Also, lathes make precise pistons for cars. 4-4) Car Engine parts must be precisely flat.

4-5) Our world moves with gears made by wedge-shaped tools.

4-6) Electricity powers, wedge tipped Mill tools that shape airplane parts like these wing
spars.

4-7) Tools have to be carefully controlled to make quantities of quality parts.

2 - Cutting Wedges
-Axes to Airplanes   VIDEO!- Script

5-1) FIVE, Controlled Cutters
At first, skilled people control the machine tools to make parts.
5-2) Next, people invent motors to move tools. They are controlled by numbers and math. People program these early machines with punch tapes.
5-3) When computers are improved, people write software to control the machine tools.
5-4) Today, computer controlled machines have many wedge-shaped cutting tools.
5-5) Simply said, withOUT wedge tools, there will be no cars or most of the every day objects we use.
5-6) People are currently working on ways for computer machine tools to self-learn how to make parts.
5-7) Next, more wedges. 6-1) SIX, Everyday Wedges
Wedge shaped objects are all around us.
6-2) Our teeth are wedges. Gear teeth are too.
6-3) These garden, hand tools are wedges.
6-4) Can openers, scissors and zippers use wedges too.
6-5) Like water pumps, wedge-shaped ship propellers push water backwards. This pushes ships forward.
6-6) Our complex High Tech world has objects made by simple, wedge-shaped tools.
6-7) Which leads us to wedges on wings.
7-1) SEVEN, Airplane Edges
The first planes have propellers that are twisting, turning wedges.
They cut through the air and push it back to thrust the plane forwards.
7-2) Propeller or prop planes have flat straight wings.
Wings have wedge-shaped cross-sections. This causes air pressure to push the plane up. This is called lift.
7-3-1) Today, Jet Engines have fan blades. Can you guess their shape? Yup! Wedges. They push together or compress air into the engine. More air or oxygen causes the fuel to burn better.
7-3-2) Why do jets have wedge shaped wings?
7-4-1) Powerful jet engines will break flat wings.
7-4-2) Jet Engines are so powerful that they would make flat wings fall apart.
7-5) Thankfully, air flows smoothly over the swept back, wedge-shaped wings with jet engines.
7-6) People and computers steer planes with parts that have wedge-shaped cross-sections.
7-7) On Airplanes, the important front face is called the "Leading Edge". The idiom, "Cutting or Leading Edge" means the latest in technology. All this starts at wedge-tips.
C-1) To Close, Ancient people use sharp rock wedges for axes and arrows.
C-2) Next, sharp metal tools cut wood for our homes, carts and ships.
C-3) Later, stronger tools shape metal parts too.
C-4) These are examples of wedges that are around me.
C-5) Today, computers control machines with wedge tools that make many of my high tech objects.
C-6) Wedge shaped tools make many airplane parts too. Jet planes are even wedge shaped.
C-7) When we know about wedges, we better understand the science inside the Cutting Edge Tech of our world today!

2) Airplanes — Wedge Tools

Cave people make wedge-shaped tools to catch and cook their food. Throughout time, people learn more science and progress from stone to metal and then powered tools. Today, people program computer-controlled V-shaped tools to make many of our everyday objects.

EDUSTORE AFRICA

Inde Ed Project
Charitable Orgn

Wedges

Wing Ways
— Self Control

Written by Douglas J. Alford
and Sally Kimangu
Illustrated by Jim Wilde

Wing Ways

Choose Wise Choices

Self Control

Choices
like feathers
impact
where we fly.

Wing Ways
— Self Control

Table of Contents

Fa nervously sits near the nest.

Today, is the day! After
weeks of pre-flight studies
and preparations, he is going
to try to fly. Fa excitedly
shifts his feet side to side.

1

With a gentle jump, Fa glides from
his branch down to another branch.

Dad said, "Good first glide Fa,
but there is more to learn.
Watch the wingless squirrel."
Fa sees the squirrel jump and glide.

Dad replies, "Watch this please."
Dad jumps. He flaps and flies up. He
flies left and right then side to side.
Next, Dad flies up and then down.

Dad lands on a branch near Fa.
"Flying is more than gliding on the
wind and gravity pulling you down. It
is about self control. You chose where
you want to go. Control your own
feathers to get you there and back."

Flying is all about four forces and three turns.

a) Gravity

Dad grabs an apple in his beak.
He lets it go. The apple falls to the
ground. "Gravity pulls everything
down. That is, a big object pulls
smaller ones towards it. When we
leave the branch, gravity pulls us
down. So, if gravity pulls us down,
we need something to push us up."

7

"First let's talk about
our feathers and wings."

It is all about air flow. When we
flap, we push air down, this pushes
us up and keeps us in the air.

b) Lift

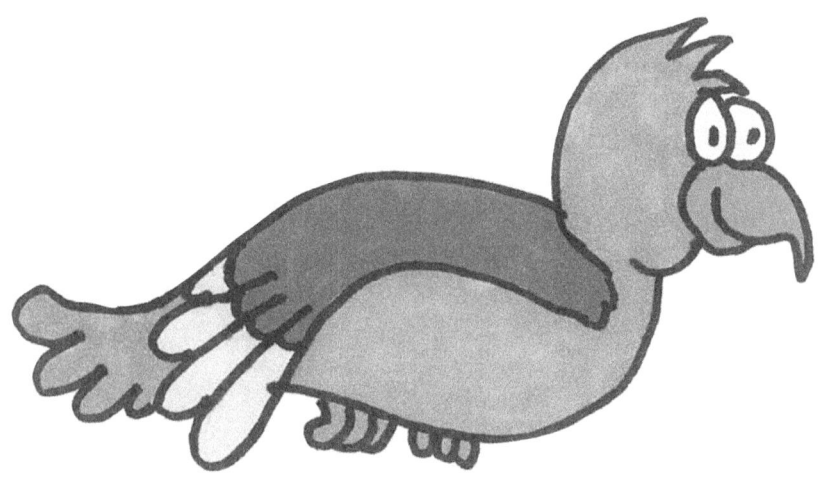

"See the curved shape of my wing. Air flows faster over the top and then pushes down. This lifts us up. Here is another way to think about this. There is more pressure in the air below. So, the air under pushes us up!"

c) Thrust

"The big feathers at the
end of our wings flap and
push air back too. This
thrust moves us forward."

d) Drag

"We birds have a special shape.
It is called streamline.
See the bike rider on the road
below. See the streamlined shape.
This makes air flow smoothly over the
rider. Air like water, flows. When we
move through air, air pushes back a bit
on us too. This is called drag."

"Our streamlined shape helps us fly easier through the air."

Dad reviews.
"Lift up, to overcome the pull down of gravity. Thrust forward to overcome the drag back from the air in front."

13

"Wow," said Fa, "We fly with
four forces." Fa practices:
up, down, forward and back.

Flight Controls

Dad teaches, "Remember, there is more to do before you fly by yourself!" Dad smiles, "Birds came first before airplanes. To help understand controls, let's look at how airplanes fly! Three different flaps control the plane."

15

1) Left or Right

This moves the plane left or right.

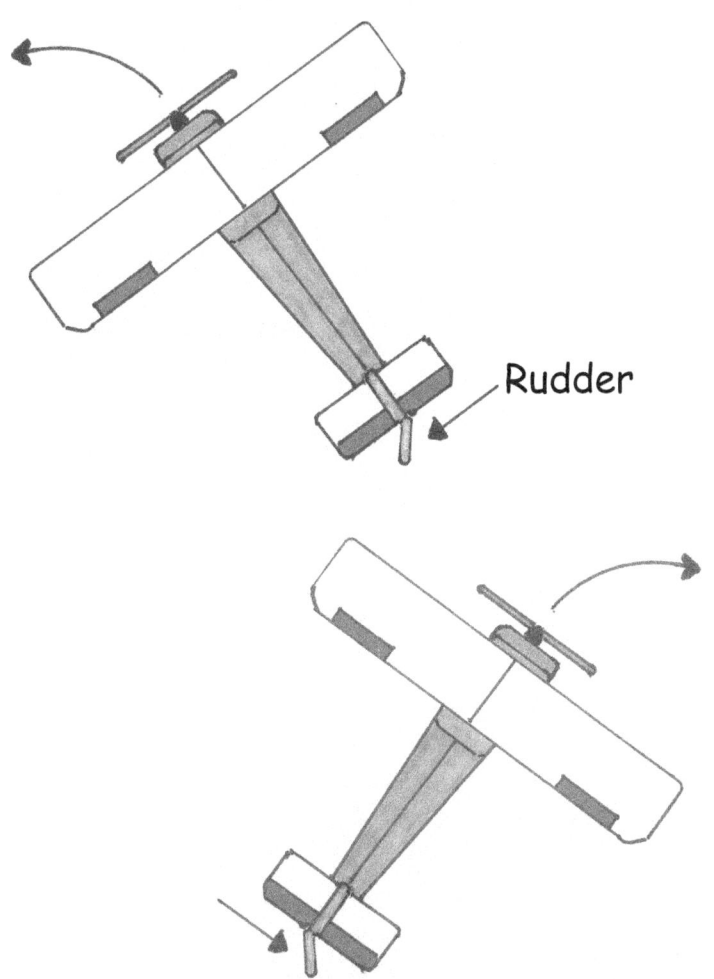

Rudder

2) Up or Down

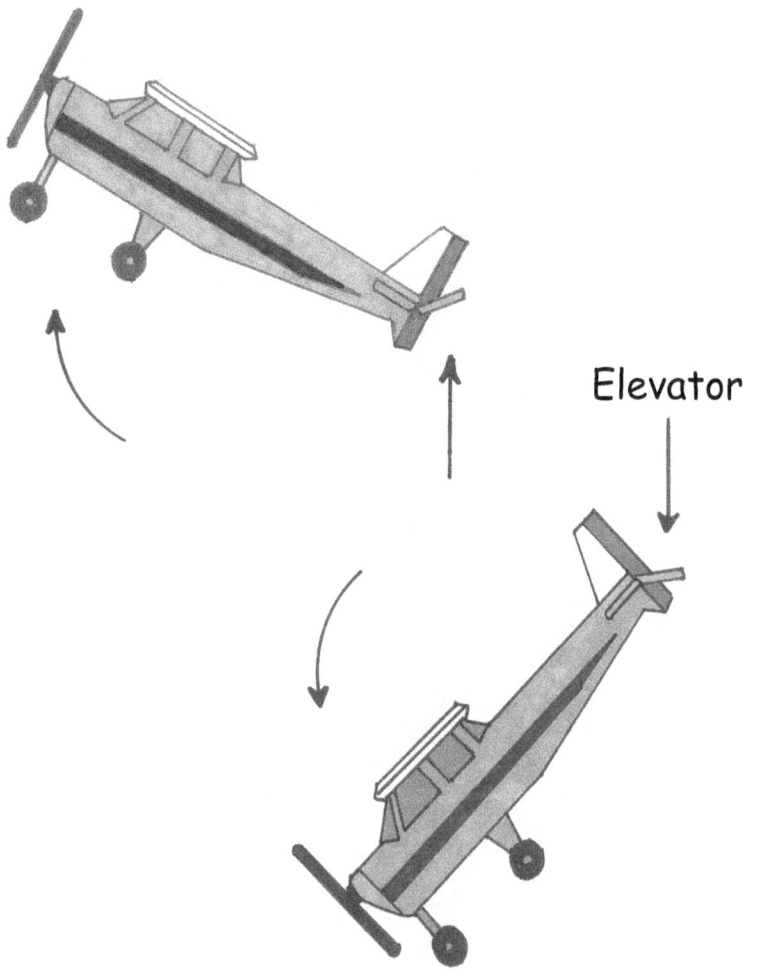

Elevator

This moves the plane up or down.

3) Side to Side

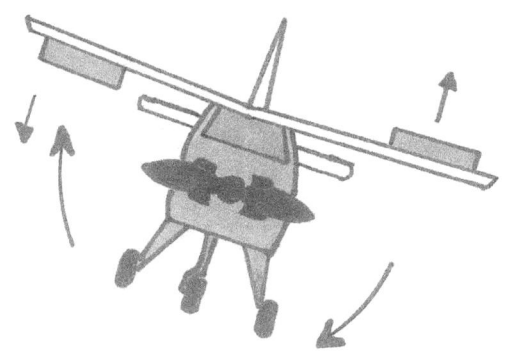

Aileron

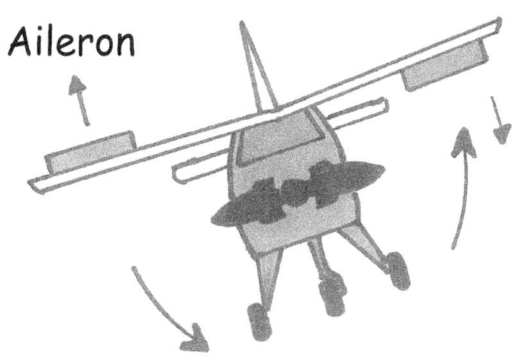

These flaps in the wings
move the plane side to side.

Similar to plane flaps, we
move feathers in our tail
and wings to control where
we go. It takes practice.

"One more important point.
How do we slow down and stop?"

"Changing the shape of our wings helps air slow us down. Remember, air pushes back or drags on shapes moving through it. When we slow down, gravity then pulls us down to help us land."

Wing Ways

Self Control

Fa is amazed!
"Where I choose to
move my feathers,
results in where I fly!"

The same is true with us.
The choices we make,
result in where we go.

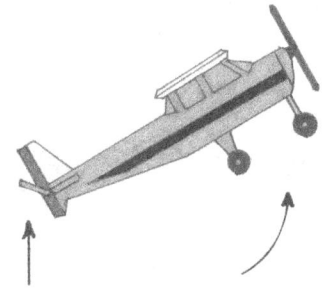

Wing Ways

Choosing choices is like learning to fly.
Our choices control the directions and
destinations of where we go in life.

More
Birds, Bugs and Bats

Animal winds do the jobs
of both airplane wings (lift)
and engines (thrust).

Airplanes fly with science!

What Is It?

The STEM-Zen Program

is an integrated SCIENCE Program with thousands of pages and over 50 videos. Teachers help students go from science empty to knowledge enLighted!

STEM-Zen Program Strategy
Integrated Science Curriculum

STEM-Zen PREP
1) Cookie Come Froms
2) Cozy Clozy
3) Good Food Goes Bad
4) No Plants No Food
5) Plants Give
6) Sand Sea
7) Senses
8) Sun Above Clouds
9) Tad's Tale
10) Too Much Tech
11) Tree Trips
12) Wing Ways

Bonus
.Cats & Dinos
.Desert Rain
. Math
  — Numbers, Money, Shapes
. Home Stars
. Teams

STEM-Zen ONE
1) Seven Ideas
2) AIRPLANES
3) CARS
4) COMPUTERS
5) Smartphone-7 Waves
6) Smartphone
   — Objects Before Apps
7) Electricity
8) Everyday Objects
9) Stress Less
10) VIDEO GAMES

STEM-Zen ONE
Teachers Guide

STEM-Zen TWO
1) Science Thinks
2) PLANES
   — Past & Present
3) WEDGE TOOLS
   — Axes to Airplanes
4) Manu-FACTORY
   — Clay to Cars
5) COMPUTERS
   — Then & Now
6) NETWORKS
   — Wires to WiFi
7) PANDEMICS
   — Causes & Cures
8) MOON RACE
   — Chase to Space
9) Science of
   Lucky Stars

STEM-Zen THREE
1) Air, Water & Food
2) BOTS
   — Automata to AI
3) POWER
   — Windows to Wheels
4) NATURE
   — Where Life Lives
5) SEASONS
   — Turn, Tilt & Orbit
6) Images in Action
   — Why Movies Move
7) LIGHT
   — Sun to Screens
8) Bright Reading
   — Baas to Books

Science
by Subject